The Theory of Particle Interactions

The Theory of Particle Interactions

V. V. Belokurov
D. V. Shirkov

Translated by
Patricia A. Millard

AIP

American Institute of Physics New York

Originally published as *Theoriia vzaimodeĭstviĭ chastits* © 1986 by Nauka.

Library of Congress Cataloging-in-Publication Data

Belokurov, V. V. (Vladimir Viktorovich)
[Theoriia vzaimodeĭstviĭ chastits. English]
Theory of particle interactions/V. V. Belokurov, D. V. Shirkov.
p. cm.--(Translation series)
Translation of: Theoriia vzaimodeĭstviĭ chastits.
Includes bibliographical references.
ISBN 0-88318-715-9
1. Nuclear reactions. I. Shirkov, D. V. (Dmitriĭ Vasil'evich) II. Title.
III. Series: Translation series (New York, N.Y.).
QC794.B43413 1990
539.7′5--dc20
89-18613
CIP

American Institute of Physics
335 East 45th Street
New York, NY 10017

Contents

Chapter 5. Theory of the electroweak interaction

Chapter 6. Quantum chromodynamics

Chapter 7. Hypothetical models of interactions

Preface to the American Edition

The theory of quantum fields, which provides the framework for a theoretical description of microparticles and their interactions, is the most advanced chapter of a modern theoretical physics.

The key element of this theory is the notion of a quantum field. From the present viewpoint, it represents the universal form of matter, underlying all its physical manifestations. One quantum field describes all particles (and antiparticles) of a given sort in the universe. The interaction of the particles is represented by the interaction of several fields of one or several species at a common space-time point.

The way quantum fields interact is determined by the symmetries existing in nature. This, the so-called gauge dynamics principle and its realizations for the known types of interactions, are explained in detail in the book. The gauge principle essentially simplifies the inner generalizations of the theory which are also discussed in the book.

The ideas and results of quantum field theory play more increasing roles in other parts of physics, such as astrophysics, cosmology, and nuclear physics. Also, many theoretical methods that had been created there were successively applied to the other diverse parts of theoretical physics. In this context, quantum field theory is a frontier part, both physically and methodologically, of the whole of theoretical physics.

This book can be recommended as the first reading for those students and physicists of various specializations to comprehend the ideas and the main methods and results of modern quantum field theory—the most abstract and mathematical branch of theoretical physics.

The authors also hope that the readers of this book will form a more complete notion of the results in this field obtained by Soviet physicists.

During the last several years, some more experimental proofs of the theories of fundamental interactions described in the book were obtained. Currently, the most significant theoretical results are associated with the superstring paradigm. However, they are far from being tested experimentally.

Preface to the First Edition

"To inflict learning gently, avoiding bloodshed as far as possible." It is this principle, formulated by a great Russian satirist, which we have attempted to follow in describing the historical development and current status of quantum field theory—the most abstract and mathematical branch of physics.

During the nearly half-century of its existence, the outward appearance of elementary particle theory has changed several times. Its history can conveniently be divided into three stages. The first 30 years were characterized by the progressive development of quantum electrodynamics, from the fundamental studies of the 1920s to the creation of the general theory of renormalization and the renormalization group method by the mid-1950s. The second period, which lasted about 15 years, was distinguished by the appearance of various theoretical schemes and approaches providing alternatives to local quantum field theory, whose development had stalled owing to difficulties encountered in applying it to the weak and strong interactions. At this time the terms "elementary particle theory" and "quantum field theory" had different meanings for most physicists. In the early 1970s the situation began to reverse quite rapidly. Quantum field theory, augmented by the dynamical principle of local gauge symmetry, confidently came to the fore, taking a firm position not only in electrodynamics, but also in the theories of the weak and strong interactions. The developmental spiral of the theory of particles and their interactions therefore came full circle, and in doing so reached a new qualitative level.

The rehabilitation of local quantum field theory on a new, gauge foundation, a triumph completed in the early 1980s, occurred relatively rapidly. This, together with the use of new, unobservable physical concepts (such as the quantum number "color," and quarks and gluons), hinders its assimilation by physicists of other specializations.

The present book originated in a brief series of lectures presented a number of times at several universities by one of the authors to a broad, but well-qualified audience of physicists. The original material of the lectures was expanded considerably in the writing of this book. This text should not be viewed as a detailed chronicle of the achievements of elementary particle theory. The authors want to highlight and at the same time comprehensibly sketch only the principle stages of its development, evaluated as such on the basis of their importance for understanding the current state of the theory. Therefore, the aim of the book is to help physicists who are not specialists in this field comprehend a new set of ideas. We have not even attempted to reflect all the contributions of the various scientists involved. We mention only a few of them by name, and give their initials only at the first mention.

In a text of this type it is impossible and, apparently, simply irrational to completely avoid the use of mathematical concepts and equations. However, the authors have tried to keep them at a minimum, and all concepts which might be unfamiliar to a substantial fraction of readers are accompanied by the necessary explanations.

At the end of the book we give a short list of relatively accessible literature recommended for further study of the topics discussed here.

We are grateful to many of our colleagues for their help and useful discussions. Particular mention should be made of A.M. Baldin, D.I. Kazakov, M.G. Meshcheryakov, V.I. Ogievetskiĭ, A.V. Radyushkin, A.A. Slavnov, and A.N. Tavkhelidze, whose advice was used in writing several sections. We would also like to thank the reviewer, L.B. Okun', for his careful reading of the manuscript and generous remarks and advice, most of which we have incorporated into the text.

Chapter 1
Basic concepts

1. Particles and interactions

The region and objects under study

Elementary particle physics is the study of the laws governing matter at distances of the order of and smaller than the nuclear scale ($r \lesssim 10^{-13}$ cm). The phenomena occurring at such small distances, comparable to the sizes of the particles themselves, are studied by a special method: particle collisions. Here it is necessary to accelerate the colliding particles to very high energies. This field is therefore also referred to as high-energy physics.

Particle physicists usually use the system of units in which $\hbar = c = 1$. In this system energy, momentum, and mass are measured in the same units. The units used at various energy scales are based on the electron volt (eV):

$$1 \text{ MeV} = 10^6 \text{ eV}, 1 \text{ GeV} = 10^9 \text{ eV}, 1 \text{ TeV} = 10^{12} \text{ eV}$$
$$(1 \text{ GeV} = 1.602 \times 10^{-10} \text{ J} = 1.782 \times 10^{-24} \text{ g}).$$

Similarly, lengths and times are measured in identical units. In this system the units of energy and length are not independent, but are expressed in terms of each other; in particular, $1 \text{ GeV}^{-1} \approx 2 \times 10^{-14}$ cm.

At present, a very large number of particles are known.

Particles of spin 1/2 which participate only in electromagnetic, weak, and gravitational interactions (see the end of this section) fall into a special class, that of the so-called leptons. The electron, muon, τ lepton, and their three associated neutrinos are leptons.

The hadrons—particles which also participate in the strong interaction—form a large group. Nucleons (the proton and neutron), mesons, baryons, and resonances are hadrons. The current view is that hadrons are composite particles composed of more fundamental objects called quarks.

Particles such as the photon, the intermediate W and Z bosons, and the so-called gluons act as carriers of interactions.

The "elementary" nature of particles

As higher energies become attainable experimentally, i.e., as it becomes possible to study phenomena at ever smaller distances, the very concept of what is an elementary particle changes. Of the particles listed above, the photon, the leptons, and the quarks are currently considered to be elementary (structureless). Later on we shall add a few more particles to this list—the interaction carriers: the gluons and intermediate bosons. Of course, it is not impossible that further study will reveal that the particles presently assumed to be elementary also have structure. Consequently, the terms "elementary particle" and "elementary particle physics" can be ambiguous, depending on the context. Therefore, when using these terms we shall avoid the adjective "elementary" as far as possible.

Particle characteristics

The fundamental particle characteristics are the rest mass, the spin (intrinsic angular momentum), electric charge, lifetime, and a few other quantities. The fact that a particle has a spin and a definite lifetime is a manifestation of its quantum properties. We will discuss these a bit later. Let us first recall some details about masses and charges.

The mass m, energy, and momentum of a particle are related by the familiar relativistic expression $E^2 - c^2\mathbf{p}^2 = m^2c^4$. As mentioned above, the mass and momentum are usually measured as energies. For example, the electron mass is $m_e = 9.110\times10^{-28}$ g $= 0.5110$ MeV and the proton mass is $m_p = 1.673\times10^{-24}$ g $= 0.9383$ GeV. Of the particles currently known, the intermediate W and Z bosons have the largest mass (~ 100 GeV). The masses of the photon and a few other particles are zero.

Particles also possess a set of discrete characteristics—the so-called internal quantum numbers. The most familiar of these is the electric charge, which for all the observed particles is a multiple of the electron charge $e = 4.803\times10^{-10}$ esu $= 1.602\times10^{-19}$ C.

The internal quantum numbers include baryon and lepton number, parity, and also the quark flavors—characteristics like isospin, strangeness, etc., determining the type of quark (and, accordingly, the type of hadron containing that quark). Moreover, quarks of the same flavor can differ by another quantum number called color. These characteristics will be discussed in more detail in Sec. 8.

Internal quantum numbers are introduced to formalize the regularities observed experimentally in processes involving particles. For example, the absence in nature of some process can be represented as a consequence of the existence of a certain quantum number whose conservation forbids that process.

Quantum properties of particles

Spin, the intrinsic angular momentum of a particle unrelated to its motion in space, has a quantum origin.

According to quantum mechanics, particle spins can take only discrete values—half-integer multiples of Planck's constant $\hbar = 1.054\times10^{-27}$ erg·s $= 6.582\times10^{-22}$ MeV. For example, the spin of the π and K mesons is zero, the spin of the neutron, proton, electron, muon, and neutrino is 1/2 in our units, and the spin of the photon, gluons, and W and Z bosons is 1. Particles of higher spin also exist.

The nonzero spin of a particle selects a certain direction in space. The angular characteristics of any process therefore depend on the spins of the particles involved in it. In practice, this makes it possible to determine the values of particle spins by analyzing the angular characteristics of processes.

Particles are divided into two classes depending on their spin: particles with integer (in units of $\hbar$) spin and particles with half-odd-integer (1/2, 3/2,...) spin. The laws governing the behavior of systems of particles of these two classes are different. Let us consider this in more detail. The quantum mechanical description of a system of particles is based on the principle of indistinguishability, according to which only quantum states which do not change under permutations of identical particles are realized in nature. The principle of indistinguishability requires that physical characteristics be invariant under interchanges of identical particles. From this it follows that the wave function of a system of identical particles is symmetric or antisymmetric under such an interchange, i.e., either it is not changed when the particles of any pair are interchanged, or it is multiplied by -1. The former occurs for systems of particles with integer spin, and the latter for those with half-odd-integer spin.

The requirement that the wave function of a system of identical particles with half-odd-integer spin be antisymmetric is known as the Pauli principle. When interactions are neglected, the Pauli principle implies that no more than one of a given type of particle can exist in each quantum state.

The difference between the symmetry properties of wave functions is related to the fact that systems of particles with integer and half-odd-integer spins obey different laws of statistics, referred to as Bose–Einstein and Fermi–Dirac statistics, respectively. Therefore, particles with integer spin are called bosons, and those with half-odd-integer spin are called fermions.

In Bose–Einstein statistics the statistical weight of any possible state of a system of N particles is 1. In Fermi–Dirac statistics the statistical weight can have two values: 1 if no occupation number exceeds unity, and 0 otherwise. We recall that the occupation numbers are the sets of numbers $N_1,..., N_s,...$, specifying how many particles of a given type occupy the individual states labeled $1,..., s,...$; $\Sigma N_i = N$.

The Pauli theorem gives a unique relation between the spin and the type of statistics. This theorem is fundamental, i.e., it is a direct consequence of the principles on which present-day quantum field theory is based.

Another manifestation of the quantum properties is the fact that most of the known particles live only a finite time before decaying. The term "decay," borrowed from nuclear physics, does not refer to the breakup of a particle into its constituents, but to its transformation into a set of different particles. One of the first decay processes known was that of neutron β-decay, in which the neutron is transformed into three particles: a proton, an electron, and an electron antineutrino, i.e., $n \to p + e^- + \bar{\nu}_e$.

Like any quantum process, particle decay is probabilistic in nature. Its probability depends exponentially on the time t: $w = 1 - \exp(-t/\tau)$. The parameter τ, which has the dimensions of time and characterizes the rate of this process, is referred to as the particle lifetime. Of course, like any probabilistic quantity, it is not a feature of an individual particle, but rather a certain average for a large set of particles.

Particle lifetimes differ widely. For example, for the neutron $\tau \approx 10^3$ s $\approx$ 15 min, while for the shortest-lived particles, the so-called resonances, it is 10^{-23}–10^{-24} s. A large group of unstable hadrons—mesons and baryons—have lifetimes intermediate between those of neutrons and resonances.

The difference between particle lifetimes is in part related to the strength of the interactions responsible for the decays. For example, neutron decay is caused by the weak interaction, and resonance decays by the strong interaction.

With few exceptions, almost all the known particles decay. Only the electron, proton, photon, and neutrions are absolutely stable, i.e., have practically infinite lifetime in the free state, within the limits of accuracy of the present experimental data. (We shall discuss searches for proton decay in Sec. 23.)

Interconvertibility of particles

The particle instability discussed above is a manifestation of the most important feature of the microworld—particle interconvertibility. This means that particles and sets of particles, as a rule, undergo transitions into other particles and sets of particles if such transitions are not forbidden by any conservation laws.

The interconvertibility of particles provides the basis for their interaction mechanisms. Delaying the detailed discussion of this problem until the next section, here we only note that, according to the current point of view, any particle interaction process can be represented as a set of elementary events with certain particle transformations occurring in each.

Therefore, in the microworld particle conversion processes are ordinary and necessary phenomena. Moreover, the absence of any transitions allowed by the known conservation laws is perceived as an indication that a new, as yet unknown conservation law exists.

The *conservation laws* for internal quantum numbers are, as mentioned above, a sort of formalization of experimentally observed regularities. The most familiar example is electric charge conservation. Also, all the presently existing experimental data are consistent with baryon and lepton number conservation (see Sec. 23).

At the same time, certain regularities are observed in processes of a particular type, but not in others. In such a case the conservation laws for the quantum numbers introduced are not general, but have a limited region of validity. Examples of such quantum numbers are the quark flavors, which are conserved in the strong interaction but not in the weak. Another example is parity, which is always conserved except in weak interaction processes.

As is well known, fundamental laws such as energy, momentum, and angular momentum conservation are a consequence of symmetries in space-time. By analogy, the conservation of internal quantum numbers can be associated with symmetries in certain abstract (internal) spaces. It is clear from the above discussion that most internal symmetries are approximate. Except for the symmetries associated with electric charge and baryon and lepton number conservation, and also the special quark "color" symmetry, internal symmetries are present in certain interactions and are violated in others. In the next chapter we shall discuss the role of approximate symmetries in particle physics in more detail.

Types of interactions

At the present time it is thought that all the various types of interactions between matter reduce to four types of fundamental interaction between particles: the strong, weak, electromagnetic, and gravitational interactions.

As already mentioned, the strong interaction involves a large group of particles, which are collectively referred to as hadrons. This interaction occurs at distances smaller than the characteristic nuclear scale $r_{\text{nucl}} \sim 10^{-13}$ cm, which is close to the Compton wavelength of the π meson. For example, the strong interaction is responsible for the nuclear forces binding protons and neutrons to form nuclei. This force is termed "strong" because it leads to large quantitative effects in the microworld. As a rule, the cross sections for strong interaction processes are close to the geometrical value ($\sim 10^{-26}$ cm^2) determined by the square of the interaction range and corresponding to the collision of spheres of radius 10^{-13} cm. At short distances ($\lesssim r_{\text{nucl}}$) the strong interaction is large in absolute value and considerably stronger than, for example, the electromagnetic interaction. The average binding energy of nucleons in a nucleus is of order 10 MeV per nucleon, which is one million times greater than the typical energy of the electromagnetic coupling of the outer electrons to a nucleus (~ 10 eV). Another way of comparing the interactions is to compare the dimensionless numerical characteristics describing the interaction strengths. A measure of the strength of the electromagnetic interaction is the so-called fine structure constant $\alpha = e^2/\hbar c = 1/137$. Strong interactions are traditionally characterized by the pion–nucleon interaction constant g. The numerical value of the dimensionless combination $g^2/\hbar c$ analogous to the fine structure constant is 14.7, which is several orders of magnitude greater than α. The current view is that the hadron interaction is not fundamental, but is the consequence of the elementary interaction between the quarks of which hadrons are composed. Chapter 6 is devoted to describing the theory of this quark interaction.

A typical example of the weak interaction is neutron β decay and the resulting radioactive decay of atomic nuclei. Another example is muon decay ($\mu^- \rightarrow e^- + \nu_\mu + \bar{\nu}_e$). Both hadrons and leptons participate in weak interactions. At energies $\ll 100$ GeV the interaction strength depends significantly on the characteristic energy of the process and increases as a power of the energy. At fairly low energies the interaction effects are actually quite weak. For example, at energies on the order of the nucleon mass (~ 1 GeV) the dimensionless quantity $G_F M_N^2$ characterizing the interaction strength is

10^{-5} (here G_F is the Fermi constant). However, at ~ 100 GeV weak interactions are comparable in strength to electromagnetic interactions. We shall consider these questions in more detail in Chap. 5 (Sec. 17), where we discuss the current theory of the weak and electromagnetic interactions.

Electrically charged particles participate directly in electromagnetic interactions. These interactions are responsible for the binding of atomic electrons to nuclei and the binding of atoms to form molecules.

The gravitational interaction is universal. It occurs between all forms of matter and between all particles. However, at the distances corresponding to the energies attained in particle interactions gravitational effects are negligible compared to the effects of the other interactions. In particular, the gravitational constant G_N (Newton's constant) in units where $\hbar = c = 1$ is 33 orders of magnitude smaller than the Fermi weak interaction constant G_F. Another illustration of the weakness of the gravitational force is the fact that the gravitational attraction between two protons is about 10^{37} times smaller than the strength of their electromagnetic repulsion. However, gravitational interactions dominate for macroscopic objects which overall are electrically neutral.

Both the gravitational and the electromagnetic interactions are long-range. Their strength falls off slowly at large distances ($\sim r^{-1}$), so it is these interactions which play the dominant role in macroscopic physics. Even on the atomic scale ($\sim 10^{-8}$ cm) the electromagnetic force is considerably stronger than the nuclear force.

2. Quantum fields

A powerful theoretical apparatus—quantum field theory—has been created for describing particle dynamics. The fundamental physical object in this theory is the quantum field, the unique synthesis of the concepts of the classical field as in electromagnetism and the probability field of nonrelativistic quantum mechanics. It is currently believed that this field is the most fundamental and universal form of matter on which all the physical manifestations of matter are based.

The classical field

This concept encompasses all physical quantities $u(x)$ depending on the spatial coordinates and time [x is a point in spacetime: $x = (\mathbf{x},t)$]. A special role is played by the wave fields involved in the transmission of interactions between particles. A typical example is the Faraday–Maxwell electromagnetic field describing the interaction of electrically charged particles.

We note that the field concept arises naturally when one attempts to avoid the idea of instantaneous action-at-a-distance between particles—an idea which is inconsistent with special relativity. Assuming that the space between particles is filled by a field, we assign to this field the function of transmitting a perturbation from one particle to another with finite velocity. Thus, the introduction of the classical field in physics is dictated by considerations of relativistic invariance.

Viewing the field as a mechanical system with an infinite number of degrees of freedom, we can construct the Lagrangian for this system and, using the variational principle of stationarity of the action, obtain the equation for the field function $u(x)$. Here dynamical invariants such as energy, momentum, and angular momentum are formed like the corresponding quantities in classical mechanics and are expressed in terms of $u(x)$ and its derivatives.

From the methodological point of view, instead of a continuous distribution it is useful to first consider a discrete mechanical system having a countable number of degrees of freedom, and then go to the continuum limit. This is conveniently done using the discrete momentum representation, which is obtained by confining the field to a spatial cube of volume $V = L^3$ and imposing the condition of periodicity with period L on each spatial coordinate. Then, as is well known, the Fourier integrals are replaced by series and the field will be determined by a denumerably infinite set of quantities $\varphi(\mathbf{k})$, where $\mathbf{k} = (2\pi/L)\, n_1, (2\pi/L)\, n_2, (2\pi/L)\, n_3$ and n_1, n_2, and n_3 take integer values from $-\infty$ to $+\infty$. In the case of free fields, this can be regarded as an expansion of the field in an infinite sum of independent harmonic oscillators, each characterized by its frequency $\omega(\mathbf{k})$ and amplitude $\varphi(\mathbf{k})$. The continuum limit is reached by allowing L to go to infinity.

Quantization

The quantization of a field is essentially the quantization of each of these oscillators. As is well known from quantum mechanics, the oscillator energy can take a discrete set of values:

$$E_n(\mathbf{k}) = \hbar\omega(\mathbf{k})\cdot(n + 1/2), \quad n = 0,1,\dots .$$

The uniform spacing of the energy levels, i.e., the linear dependence of the energy on the state number, provides the basis for the corpuscular treatment of the theory. An oscillator with frequency $\omega(\mathbf{k})$ in the nth excited state can be represented as a set of n indistinguishable particles—excitation quanta, each of energy $\hbar\omega(\mathbf{k})$.

The number of particlelike excitations can obviously be arbitrary, which makes it possible to construct a quantum description of a system with a variable number of particles. Here the field functions become operators which change the number of particles in the system.

More precisely, each quantum field $u_i(x)$ is written as a linear combination of creation operators $u_i^+(k)$, describing the creation of a particle of mass m_i and 4-momentum k, and annihilation operators $u_i^-(k)$, corresponding to the annihilation of such a particle. The creation operator $u^+(k)$ acting on a state with energy–momentum p containing N particles transforms it into a state with $N+1$ particles and energy–momentum $p+k$. The annihilation operator $u^{-}(k)$ decreases the energy–momentum and particle number of the state in the analogous manner.

Therefore, as a result of quantization the field functions become operators (in the quantum mechanical sense). These operators act on a wave function Φ common to all the fields. In conformity with ordinary quantum mechanics, the wave function Φ completely characterizes the physical state of a system described by quantum wave fields. Like the usual function Ψ, the function Φ can be viewed as a vector in some linear space. It is therefore referred to as a state vector. The expectation values of dynamical quantities and transition probabilities are expressed in terms of quadratic forms of Φ, so Φ is also referred to as the amplitude of the state. The vacuum state plays a special role in this formalism. Its amplitude is denoted by the symbol Φ_0. By definition, there are no particles in this state, it has zero momentum, the minimum energy, and the annihilation operator of any field applied to it gives zero:

$$u^{-}(k)\Phi_0 = 0.$$

The creation operator $u^{+}(k)$ acting on Φ_0 gives a one-particle state. A state containing N particles can be constructed by allowing N creation operators to act on the vacuum state:

$$\Phi_N = u^{+}(k_1)...u^{+}(k_N)\Phi_0.$$

When the number of particles is not fixed, the amplitude of the state is written as a linear superposition of the amplitudes Φ_N with certain weight functions:

$$\Phi = \sum_{N=0}^{\infty} \chi_N \Phi_N.$$

This is the Fock representation of the state amplitude.

Quantum fields

We have shown how the quantum field arises as a result of quantization of the classical field. For example, upon quantization the potentials $\mathbf{A}$ and φ of the classical electromagnetic field and the field strengths $\mathbf{E}$ and $\mathbf{H}$ become quantum operators which operate on the quantum mechanical wave function of the physical system. Linear combinations of these field operators are the creation operators for quanta of the electromagnetic field—photons. At the same time, the operator potentials and field strengths satisfy the usual equations of motion of electrodynamics—Maxwell's equations. In this manner we arrive at the quantum electromagnetic field, which combines the wave and corpuscular features of electromagnetism.

The concept of the quantum field can also be reached starting from particles. Quantum mechanics gives a probabilistic description of the behavior of a particle, for example, an electron, by means of a wave function satisfying the Dirac equation (see Sec. 4). The extension of these ideas to systems containing several particles of a given type (for example, electrons) is conveniently formalized by the introduction of an operator wave field written in the occupation number representation by means of creation and annihilation operators. This operator field is completely analogous to the quantum electromagnetic field. The procedure of transforming to operators of the occupation number representation is usually referred to as second quantization. It should be clear that no "second quantization" is actually involved here, so this term is not only inappropriate, but also confusing.

Therefore, the concept of quantum field contains the organic synthesis of the wave and corpuscular properties of matter. The quantum field should be viewed as a single fundamental object replacing the fields and particles of classic physics. One such field specified at each point in four-dimensional spacetime describes all the particles of a given type in the universe.

The interaction of quantum fields

In order to obtain linear equations of motion (the usual type of equation for free fields), we must use a Lagrangian, i.e., density of the Lagrange function, which is bilinear in the field functions and their derivatives. When interactions are present the equations are no longer linear, and the Lagrangian contains products of three and more fields. In a local quantum field theory it is assumed that the field functions involved in such products are evaluated at a single point in spacetime.

Since the field functions are expressed in terms of creation and annihilation operators, the elementary event in any interaction is the creation of certain particles and the annihilation of others at the same point. This appears as a local and instantaneous transformation of certain particles into others.

At this point we must make the following important observation. In classical physics a free particle can neither emit nor absorb another particle while remaining unchanged (for example, in classical physics a free electron can neither emit nor absorb a photon), since the laws of energy and momentum conservation will not be simultaneously satisfied in such processes. The situation is different in quantum theory. According to the uncertainty principle, the time interval Δt and spatial interval Δx in which a particle is located are related to the spreads in the possible energies ΔE and momenta Δp_x by the inequalities $\Delta t \cdot \Delta E \geqslant \hbar$ and $\Delta x \cdot \Delta p_x \geqslant \hbar$. Therefore, if the emitted particle exists for a small time interval, the usual relativistic relation between its energy, momentum, and mass will not be satisfied. Such particles are encountered only in intermediate states and are called virtual particles. It is easily verified that their emission or absorption is consistent with the laws of energy and momentum conservation. In particular, a free electron can emit a virtual photon whose energy and momentum satisfy the relation $E^2 - \mathbf{q}^2 < 0$.

We note that the uncertainty relations have another important consequence. Although the interaction is assumed to be local, it is impossible to specify the actual point in spacetime at which an elementary event occurs, since the energy–momentum transfer is uniquely determined by the conservation laws.

Elementary interaction events are conveniently depicted as vertices with three or more lines. For example, in quantum electrodynamics a vertex has the form:

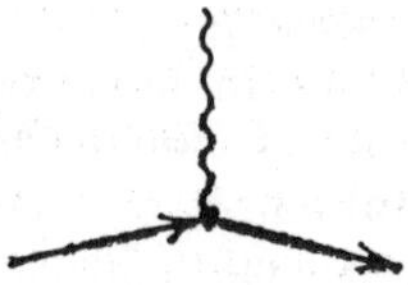

where the wavy line represents the creation (or annihilation) of a photon and the straight lines represent the annihilation of an electron in one state and creation of an electron in another.

As far as the ordinary classical forces acting between particles are concerned, they are generally represented as a secondary effect arising from the exchange of virtual particles—field quanta which transmit the given interaction. For example, the interaction between two electrons can occur as shown in the following figure:

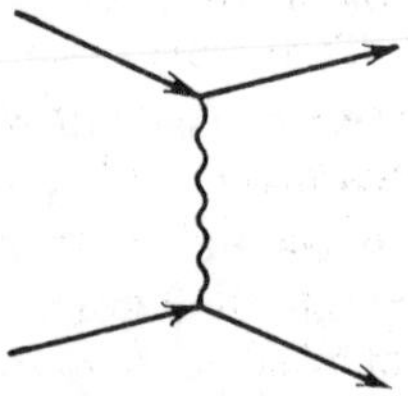

Here we should note that in some quantum field models the elementary interaction events may also describe real processes. For example, in the weak interaction Lagrangian written down by Fermi, which will be discussed in more detail in Sec. 4, the elementary event is assumed to be the actual process of neutron β decay:

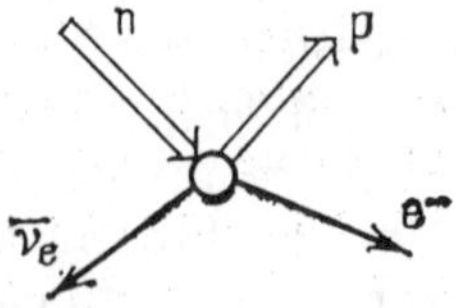

Quantum field interactions have an important feature which follows from relativistic invariance. As noted above, the quantum field functions are linear combinations of creation and annihilation operators. Therefore, if an interaction Lagrangian describes a process in which some particles are annihilated and others are created, the same Lagrangian necessarily leads to processes in which the annihilated and created particles reverse roles. An example is the process of inverse β decay: $\bar{\nu}_e + p \to n + e^+$, which is described by the same Fermi Lagrangian.

3. Symmetries and transformation groups

Spacetime and internal symmetries

An object is said to possess a symmetry if there exist transformations which leave it unchanged. The concept of symmetry is central to physics in general, and particularly to particle physics. The most important examples of symmetry are related to very general features of spacetime. Foremost is the uniformity of space and time, which is manifested in the invariance of physical laws under shifts (translations) of the spacetime coordinates. Another important feature is the isotropy of space, which implies invariance under spatial rotations. Yet another example is the equivalence of coordinate systems moving relative to each other with constant velocity. This corresponds to invariance under transformations of the time and space coordinates which leave the combination $c^2t^2 - x^2 - y^2 - z^2$ unchanged—the so-called Lorentz transformations.

These symmetries are inherent in all the fundamental interactions. In addition, there are also spacetime symmetries which occur for restricted sets of phenomena. For example, all processes except weak interaction ones are symmetric under the reflection of the three space axes—mirror symmetry (the transformation is denoted by the letter P)—and reversal of the time direction (the T transformation). There are also symmetries which are completely unrelated to the properties of spacetime. For example, in most processes particles and antiparticles behave identically. In such cases there is invariance under the exchange of particles and antiparticles (the C transformation). A few other symmetries can also be present in particle interactions. Again, each is conveniently associated with an invariance under transformations in some formal (so-called internal) space which is specified by "internal" degrees of freedom—quantum numbers which distinguish the particles in question. (A typical example is isotopic symmetry, which is discussed in Sec. 6.)

These transformations have a number of common features, which allow them to be described within the framework of a single mathematical theory—group theory.

Group theoretical concepts and methods which are relatively simple in the logical sense are widely used in contemporary particle physics. Therefore, in this section, which is included to make the later discussion easier to understand, we recall some basic definitions and also the simplest properties of groups which are encountered in later chapters. Here we shall avoid, as far as possible, extreme rigor and abstraction and shall consider only groups of transformations of vector spaces.

We shall use the following sets of transformations as examples illustrating the basic definitions:

- translations along an axis;
- rotations of the plane about a given point, the coordinate origin;
- rotations of three-dimensional space;
- reflections of the coordinates (the P and T transformations),

and also a few other transformations which we shall encounter later. It will become clear from our discussion that to this set of transformations we need to add the identity transformation, which changes nothing.

The definition of a group. Subgroups

First let us define the concept of a group itself. A set of transformations G is said to form a group if the following requirements (group axioms) are satisfied.

(1) Any two transformations g_1 and g_2 can be performed in succession, and the result, referred to as a composition of transformations or a product and denoted as $g_2 \cdot g_1$, must correspond to the result of a third transformation g_3 from the same set G, so we can write $g_2 \cdot g_1 = g_3$.

(2) The multiplication (composition) of more than two transformations is associative, i.e., $g_3 \cdot (g_2 \cdot g_1) = (g_3 \cdot g_2) \cdot g_1$.

(3) Among the transformations of the set G there exists the identity transformation, which is usually denoted by the symbol E and referred to as the identity element of the group.

(4) For any transformation g of the group G there exists the inverse transformation, denoted as g^{-1}, i.e., the transformation such that $g^{-1} \cdot g = E$.

These properties are almost obvious in the examples we have chosen.

Let us make a few additional remarks. For translations and rotations the identity transformations are the translation by the zero vector and the rotation by zero angle. The inverse transformations are translations (rotations) by the same distance (angle), but in the opposite direction. Reflection coincides with its inverse, i.e., the group of reflections consists of only the two elements P and $P^2 = E$.

If some set of transformations H of the group G itself forms a group, then H is called a subgroup of G. An example of a subgroup is the group obtained by restricting translations to those by vectors proportional to a certain vector with integer proportionality coefficients. Another example is rotations of the plane by angles equal to $2\pi r$ rad, where r is a rational fraction. Rotations about one axis form a subgroup of the three-dimensional rotation group.

Abelian and non-Abelian groups

We note that the group multiplication properties listed above do not assume commutativity, i.e., in the group definition there is no requirement that $g_1 \cdot g_2 = g_2 \cdot g_1$. Therefore, groups can exist for which this condition is not satisfied, i.e., $g_1 \cdot g_2 \neq g_2 \cdot g_1$, and such groups do actually exist. The simplest example is the group of rotations of three-dimensional space. It is easy to verify that, for example, rotations about two mutually perpendicular axes do not, in general, commute with each other.

If the multiplication of any two transformations of a given group G is commutative, this group is termed Abelian. If the group G contains at least one pair of transformations which do not commute, the group is termed non-Abelian. Of course, even in non-Abelian groups there exist some transformations which commute with each other. For example, the identity transformation commutes with all the transformations of the group. The translation group and the group of rotations in a plane are obviously Abelian, while the group of rotations of three-dimensional space is non-Abelian.

Lie groups

In contrast to the group of axis reflections, the translation and rotation groups have a very important property. Any transformation from these groups can be specified in a continuous manner (parametrized) by means of a real number or a set of real numbers (parameters). For example, any shift along an axis is specified by the length of the shift with " + " sign if the shift is in the positive direction and " − " if in the opposite. The composition of two shifts corresponds to the algebraic sum of the parameters defined in this manner. The rotation of a plane can be specified by the rotation angle $\varphi\,(0 \leqslant \varphi < 2\pi)$. Here the group operation corresponds to the addition of rotation angles modulo 2π.

The group of rotations in three-dimensional space is often parametrized by the three Euler angles φ, θ, ψ describing the positions of the initial and rotated coordinate systems. Alternatively, three-dimensional rotations can be parametrized by specifying the direction of the axis about which the rotation is made and the rotation angle. It is easy to see that an arbitrary direction in three-dimensional space is determined by two real numbers: the "parallel" and "meridian." Therefore, the total number of parameters here is again three.

The minimum number n of real parameters needed to specify any transformation of a given group G is referred to as the group dimension, and the group G itself is referred to as an n-parameter group.

Therefore, the group of rotations in the plane is a one-parameter group, and the group of rotations in three-dimensional space is a three-parameter group.

We have now arrived at the extremely important concept of a Lie group. A Lie group is a group whose parameter space can locally be identified with a finite-dimensional Euclidean space. This can be stated more rigorously as follows. A continuous group G is referred to as an n-parameter Lie group if some neighborhood of any element can be uniquely and continuously mapped into some region of n-dimensional real Euclidean space. Practically all the continuous groups encountered in particle theory are Lie groups.

The group U(1)

The simplest physically interesting example of a Lie group is the Abelian group of rotations in a plane. We have already mentioned its parametrization using points in the interval $[0,2\pi]$. We obtain another parametrization by associating with a rotation in a plane by an angle φ a number in the complex plane $e^{i\varphi}$ with unit modulus and argument φ. Here the group operation corresponds to the ordinary multiplication of complex numbers and the inverse element corresponds to the complex conjugate $(e^{i\varphi})^{-1} = e^{-i\varphi}$.

The latter equation is the simplest example of an important property called unitarity. The concept of unitarity is somewhat more complicated for matrix groups. We shall discuss this later. The group of complex numbers of modulus 1, or the equivalent group of rotations in a plane by the angle φ, is referred to as U(1), where the letter U indicates that the group is unitary. In particle physics the group U(1) arises as the Abelian group of gauge transformations associated with the conservation of additive charges.

Matrix representations

Let us now consider the more complicated groups which we shall encounter below. For this we take an arbitrary vector space of finite dimension and consider all the linear transformations which have an inverse. As is well known, to each such transformation here corresponds a unique square matrix with nonzero determinant. If the group operation is taken to be matrix multiplication, then, as is easily checked, the set of all such matrices (i.e., all linear nondegenerate transformations of the finite-dimensional vector space) forms a group. We recall that matrix multiplication is associative, but not, in general, commutative. Therefore, matrix groups are, as a rule, non-Abelian.

Unitary matrices are of special interest from the physical point of view. We recall that a matrix U is termed unitary if its Hermitian conjugate (i.e., the transpose of the complex conjugate matrix) is its inverse, i.e., $U^+U = UU^+ = \mathbb{1}$, or $U^+ = U^{-1}$. Unitary transformations are special because they preserve the squared modulus of the state vector (the wave function) $|\psi|^2$ and therefore lead to a physically equivalent picture.

It can easily be shown that the multiplication of two unitary matrices gives a unitary matrix and that a matrix which is the inverse of a unitary matrix is also unitary. The set of unitary matrices therefore forms a group.

Special unitary groups

If the elements of a matrix group are unimodular, i.e., have unit determinant, the group is referred to as special; this is denoted by the letter S. For example, the SU(N) is the group of linear transformations of a complex N-dimensional vector space, effected by unitary $N\times N$ matrices with determinant 1.

The simplest group of this series is the group SU(2) consisting of unitary, unimodular, complex 2×2 matrices. It is easy to find the dimension of this group. For this we consider an arbitrary 2×2 matrix.

This matrix is determined by four complex or eight real numbers. The requirement of unitarity imposes four conditions on these, and that of unit determinant gives another. We are therefore left with three independent real numbers for parametrizing any matrix of the group SU(2). In other words, the dimension of SU(2) is 3. In a completely analogous manner it can be shown that the dimension of SU(N) is $2N^2 - N^2 - 1 = N^2 - 1$.

Let us show for the example of SU(2) how an arbitrary element of a group can be represented. First we consider matrices infinitesimally differing from the unit matrix. Each such matrix is specified by three infinitesimal real parameters α_k ($k = 1,2,3$) and can be written as $g = \mathbb{1} + i\Sigma_{k=1}^{3}\alpha_k t_k$, where t_k are linearly independent matrices forming a basis in the neighborhood of the group identity. These are the group generators. It is easily seen that the number of generators is equal to the dimension of the group: 3 for SU(2), 8 for SU(3), 24 for SU(5), and so on.

Before considering the specific form of the matrices t_k, let us write down an arbitrary element g of the group SU(2). In group theory it has been proved that an arbitrary element of this group can be written as $g = \exp\{i\Sigma_{k=1}^{3}\alpha_k t_k\}$. Here α_k are arbitrary (not necessarily small) real parameters.

It can also be shown that for g to be unitary it is necessary that the matrices t_k be Hermitian, and for $\det g = 1$ we must have $\mathrm{Tr}\, t_k = 0$.

All the above is also valid for an arbitrary group SU(N). In this case the index k takes values from 1 to $N^2 - 1$.

Representations

We also need to introduce the following important concept. Let a group G consist of transformations of some space. We assume that there exists a set of transformations of another (different from the first) space which can uniquely be associated with the group G such that all the group properties are preserved, i.e., the product of two transformations in one space corresponds to the product of the two associated transformations in the other space, with the analogous requirement for the identity and inverse transformations. In this case the group G is realized as transformations of different spaces, i.e., it is said that it has different representations. The dimension of the corresponding space is referred to as the dimension of the representation. In order to obtain the specific form of the matrices t_k it is necessary to specify the group representation. The simplest representation of any group is the so-called singlet representation. The space in which this representation acts consists of a single element which does not change under any transformation.

If a unique correspondence can be established between the elements of a group and the transformations forming its representation, this representation is referred to as faithful. It fully reproduces the structure of the group. In particular, the fundamental and adjoint representations of the group SU(N) introduced below are faithful.

The representation of the group SU(N) used to define this group is referred to as the fundamental representation.

For SU(2) the fundamental representation is that in the form of unitary, unimodular 2×2 matrices. The generators of the fundamental representation of SU(2) are the familiar Pauli matrices $\tau_1 = \begin{pmatrix} 0 & 1 \\ 1 & 0 \end{pmatrix}$, $\tau_2 = \begin{pmatrix} 0 & -i \\ i & 0 \end{pmatrix}$, $\tau_3 = \begin{pmatrix} 1 & 0 \\ 0 & -1 \end{pmatrix}$. The generators of the fundamental representation of SU(3) are the 3×3 Gell-Mann matrices. Their form is not very important for us, so we shall not give it here. We only note that three of the eight Gell-Mann matrices are written as

$$\begin{pmatrix} \|\tau_k\| & & 0 \\ & & 0 \\ 0 & 0 & 0 \end{pmatrix},$$

where τ_k are the Pauli matrices. This is a reflection of the fact that SU(2) is a subgroup of SU(3).

An important role is played by the adjoint representation—the representation in the form of matrices acting in a real space whose dimension coincides with the group dimension.

The adjoint representation of SU(2) is the set of rotations of ordinary three-dimensional space. Its generators have the form $(t_k)_{ij} = i\varepsilon_{ijk}$, where i, j, $k = 1,2,3$ and ε_{ijk} is the totally antisymmetric unit tensor.

In the next chapter we shall become acquainted with approximate symmetries relating different particles possessing some set of common features. Here sets of particles with identical features—called multiplets—determine the representation spaces of the symmetry group. The dimension of the multiplet is the dimension of the corresponding space.

Later we will also encounter the direct (orthogonal) sum and the direct (tensor) product of representations of a group, which have dimensions, respectively, equal to the sum and product of the dimensions of the original representations. To illustrate the concept of the direct sum of two spaces, let us consider the orthogonal addition of two one-dimensional spaces (axes). The result of this is a two-dimensional space (a plane). The direct sum of a plane and an axis is a three-dimensional space.

The matrices of a representation which is the direct sum of representations are block-diagonal. The matrix elements outside the blocks lying on the diagonal are zero. It has been proved that any representation of the group SU(N) can be expanded into the orthogonal sum of simpler representations of lower dimension.

If all the matrices of a representation of a unitary group have a block-diagonal form in a suitable basis, the representation is referred to as reducible. Otherwise, the representation is irreducible. Irreducible representations of the group SU(2) can have any dimension. In contrast, the dimensions of the irreducible representations of the group SU(3) take only the value 1, 3, 8, 10,... .

The tensor multiplication of representations is more complicated. For each matrix of the first representation and each matrix of the second, one constructs the matrix consisting of all possible products of the elements of the first matrix and those of the second. Clearly, here the dimensions of the representations multiply. We shall encounter the tensor product of representations of SU(3) and its expansion into the direct sum of irreducible representations in Sec. 8 when discussing composite models of hadrons.

Let us give another, more intuitive definition of the direct product of two groups. Let G_1 and G_2 be groups of $k\times k$ and $m\times m$ matrices, respectively. Then the group $G = G_1\times G_2$ is the group of all $(k+m)\times(k+m)$ matrices of the form:

$$\begin{pmatrix} \|G_1\| & \vdots\vdots \\ \vdots\vdots & \|G_2\| \end{pmatrix},$$

where the dots denote zeros. The representation spaces of the group G are the direct sums of the representation spaces of the groups G_1 and G_2.

We conclude this section by listing the subgroups of the simplest unitary groups:

SU(2)$\supset$U(1);

SU(3)$\supset$SU(2); SU(2)$\times$U(1);

SU(4)$\supset$SU(3)$\times$U(1); SU(2)$\times$SU(2); SU(2)$\times$SU(2)$\times$U(1);

SU(5)$\supset$SU(4)$\times$U(1); SU(3)$\times$SU(2)$\times$U(1).

Chapter 2
Stages in the development of the fundamental ideas

4. Laying the foundation

By the end of the 1920s quantum mechanics had been formulated definitively and universally accepted. As is well known, this theory explained the laws of thermal radiation, the stability of atoms, and the fact that each element is characterized by its own narrow spectral lines. It allowed the sizes and excitation energies of atoms to be calculated and also led to the solution of many other problems in atomic physics.

The success of quantum mechanics in describing atomic phenomena seemed to indicate that quantum ideas should apply in the physics of the microworld. During the late 1920s and early 1930s the union of these ideas with the requirement of relativistic invariance led to laying the foundation of quantum field theory.

The relativization of quantum mechanics

As is well known, nonrelativistic quantum mechanics has a limited region of applicability. From the very beginning attempts were made to generalize this theory by making it relativistically invariant.

In 1926 O. Klein, V.A. Fock, and W. Gordon proposed a relativistically invariant generalization of the Schrödinger equation for the wave function of a free particle. This equation has the form:

$$\hbar^2 \frac{\partial^2 \psi}{\partial t^2} = \hbar^2 c^2 \left(\frac{\partial^2 \psi}{\partial x^2} + \frac{\partial^2 \psi}{\partial y^2} + \frac{\partial^2 \psi}{\partial z^2} \right) - m^2 c^4 \psi.$$

If we associate the energy and momentum with the operators differentiating with respect to the time and space coordinates, we easily see that this equation corresponds to the relativistic relation between energy and momentum: $E^2 = \mathbf{p}^2 c^2 + m^2 c^4$.

The Klein–Gordon–Fock equation does not include the additional degree of freedom of a particle associated with its spin. Because of this, for a long time—until the discovery of spin-0 particles—it was assumed that this equation had no connection to the real world.

Moreover, if we view this equation as the equation for the wave function of a particle, then, as shown by Dirac, we unavoidably come into contradiction with the probabilistic interpretation of the wave function. It can easily be shown that, since the Klein–Gordon–Fock equation contains a second-order time derivative, the expression for the probability density $(\overset{*}{\psi}\,\partial/\partial t\psi - \partial\overset{*}{\psi}/\partial t\,\psi)$ is not a positive definite quantity. This difficulty persisted until 1934, when W. Pauli and V. Weisskopf correctly interpreted this equation as the equation for a scalar quantum field.

In 1928 P.A.M. Dirac proposed his remarkable equation for the description of the wave functions of spin-1/2 particles. This equation is first order in $\partial/\partial t$ and is satisfied by a multicomponent wave function. For particles of nonzero mass ψ is a four-component column vector:

$$\psi = \begin{pmatrix} \psi_1 \\ \psi_2 \\ \psi_3 \\ \psi_4 \end{pmatrix}.$$

In other words, the Dirac equation is essentially a system of four equations for four functions ψ_1, ψ_2, ψ_3, and ψ_4 depending on the coordinates x_1, x_2, x_3 and the time $t = x_0/c$. It can be written in matrix form as

$$i\left(\gamma^0 \frac{\partial}{\partial x^0} - \gamma^1 \frac{\partial}{\partial x^1} - \gamma^2 \frac{\partial}{\partial x^2} - \gamma^3 \frac{\partial}{\partial x^3}\right)\psi = \frac{mc}{\hbar}\,\psi.$$

The Dirac matrices $\gamma^0, \gamma^1, \gamma^2$, and γ^3 are 4×4 matrices satisfying the condition:

$$\gamma^\mu\gamma^\nu + \gamma^\nu\gamma^\mu = 2g^{\mu\nu},$$
$$\mu, \nu = 0, 1, 2, 3; \quad g^{\mu\nu} = \mathrm{diag}(1, -1, -1, -1).$$

The Dirac equation is easily generalized to the case of a particle moving in an external electromagnetic field. For this we need only recall that the operators differentiating with respect to the time and space coordinates correspond to the energy E and momentum $\mathbf{P}$ of the particle and, as in classical electrodynamics, replace $\mathbf{P}$ by $\mathbf{p} - (e/c)\mathbf{A}$ and add the term eA_0 to the expression for the energy. The solution of the equation generalized in this manner for an electron moving in the field of a nucleus led to a more accurate equation for the energy levels of hydrogenlike atoms: one which describes the so-called fine splitting of levels with the same principal quantum number n and different values of the total angular momentum.

The quantum theory of radiation

Ordinary quantum mechanics is already insufficient for solving problems involving the interaction of the electromagnetic field with nonrelativistic objects, for example, the emission and absorption of light by atoms. It is necessary to have a formalism which also allows the electromagnetic field to be described quantum mechanically. This problem was solved in 1927 by Dirac, who represented the electromagnetic field as a set of quantum oscillators and included its interaction with atoms using first-order perturbation theory.

This approach made it possible to understand the nature of spontaneous emission by atoms, i.e., radiation accompanied by the spontaneous transition

of the atom from an excited state to a lower state in a vacuum, and to calculate the probability for this transition.

In a subsequent study published the same year, Dirac used his method to rederive a series of known expressions for the scattering of light by atomic systems. After this, the Dirac theory was applied to other problems involving radiation processes. An example is the work of Weisskopf and Wigner, who used it to obtain an expression for the natural width of spectral lines.

In the work of Dirac the entire electromagnetic field was not quantized, only its transverse, "radiative" part was. Electrostatic fields were included in the form of a Coulomb interaction energy in the particle Hamiltonian. This lopsided treatment of the two field terms is not invariant under Lorentz transformations, which mix these terms. The problem of constructing a relativistically invariant scheme for quantizing the electromagnetic field therefore remained acute. P. Jordan, Pauli, W. Heisenberg, and E. Fermi all worked on its solution. In the Fermi approach, which corresponds most closely in spirit with the later theory, the unphysical degrees of freedom are eliminated by imposing auxiliary conditions on the allowed state vectors. Dirac, Fock, and B. Podolsky later published studies devoted to the generalization of the theory and presented a proof of the relativistic invariance of the quantization scheme.

The study by N. Bohr and L. Rosenfeld published in 1933 played an important role in revealing the physical meaning of the quantized radiation field. They showed that the electric and magnetic field strengths obey uncertainty relations similar to the Heisenberg relations between the coordinates and momenta. Thus, for example, it is impossible to accurately measure the x component of the electric field and the y and z components of the magnetic field simultaneously.

Antiparticles

In the quantum theory of radiation electrons and photons are treated asymmetrically. Whereas photons arise in the quantization of the electromagnetic field, electrons are described by a wave function in configuration space satisfying the Dirac equation.

Meanwhile, the interpretation of the Dirac equation as the equation for the wave function of a single particle encountered a serious difficulty owing to the existence of states with negative energy. Such states are present among the solutions of the Dirac equation and cannot be eliminated from the theory. If taken literally, they would lead to absurd consequences, for example, instability of the hydrogen atom, since radiative transitions from ordinary states to states of negative energy would be allowed.

To avoid these contradictions, in 1930 Dirac made an ingenious hypothesis based on a new interpretation of the concept of the vacuum. He assumed that in the vacuum all states of negative energy are filled by electrons and form a physically unobservable background. Transitions of electrons from states of positive energy to states of negative energy are forbidden by the Pauli principle.

The Dirac hypothesis not only eliminated the difficulties described above, but also led to the prediction of a number of fundamentally new physical phenomena. For example, processes must occur in which an electron from the "sea" of filled negative-energy states obtains the required energy by, for example, photon absorption and makes a transition to a positive-energy state. The electron thereby becomes observable. The empty state of negative energy, a so-called hole, also becomes observable. In all processes a hole behaves like a particle with positive charge and the electron mass, i.e., it possesses the properties of an antielectron (positron). An electron can undergo the reverse transition into an empty hole, emitting a photon of the corresponding energy.

This theory therefore led to the prediction of the existence of antiparticles and also two fundamental processes unknown at that time—the creation and annihilation of a particle–antiparticle pair.

In addition to Dirac, I.E. Tamm, J. Oppenheimer, and H. Weyl also made important contributions to the development of the concept of antimatter. The existence of antiparticles was confirmed experimentally soon afterwards—in 1932 the positron was discovered.

We note that the Klein–Gordon–Fock equation also has solutions corresponding to negative values of the energy. The general reason that such states are present in a theory is that the relativistic relation $E^2 = \mathbf{p}^2c^2 + m^2c^4$ depends quadratically on the energy, and the square root is double-valued: $E = \pm\sqrt{\mathbf{p}^2c^2 + m^2c^4}$. Therefore, in the final analysis the existence of antiparticles is a consequence of relativistic invariance.

The Dirac hypothesis of the existence of an unobservable background of negative-energy states appeared rather artificial and was not subsequently confirmed. Moreover, it was completely inapplicable for particles with integer spin. However, it proved possible to solve the problem of negative-energy states in the so-called second quantization formalism.

Quantum wave fields

This formalism was developed at the end of the 1920s by Jordan, Klein, and Wigner for the quantum description of a system of N identical particles, and then generalized by Fock to systems with a variable number of particles. It is based on the use of creation and annihilation operators and the occupation number representation. Here a relativistic quantum field function is expressed as a linear combination of both creation and annihilation operators for the corresponding particles—field quanta. For example, the function $u(x)$ describes the simultaneous annihilation of a particle and creation of an antiparticle, and the conjugate function $\overset{*}{u}(x)$ describes particle creation and antiparticle annihilation.

In 1934 Heisenberg, and also Oppenheimer and W. Furry, showed that in the quantum treatment of the Dirac field, quantized taking into account the

Pauli principle, electrons and positrons are described symmetrically and there is no problem with negative energies. The Bose–Einstein quantization of the solution of the Klein–Gordon–Fock equation, done by Pauli and Weisskopf, leads to analogous results.

Therefore, the d'Alembert, Dirac, and Klein–Gordon–Fock equations should be interpreted as equations for relativistic operator field functions acting on the state vector of a system of quantum fields satisfying the Schrödinger equation. A new concept had arisen—that of the quantum field.

The formulation of quantum electrodynamics

In this new picture, light and matter (i.e., electrons and positrons) play symmetric and equal roles as interacting wave fields in 4-dimensional spacetime. The theory of the interaction of quantum electromagnetic and spinor fields is called quantum electrodynamics (QED).

In this theory, electrons, positrons, and photons are treated identically, as quanta of the corresponding fields. Of course, they are quantized differently. Certain commutation relations are valid for the spin-1 electromagnetic field, while the operators for the spin-1/2 electron field satisfy anticommutation relations. This leads to antisymmetric wave functions for fermion (electron) systems and the Pauli exclusion principle.

If the electron field and the electromagnetic field are expressed in terms of creation and annihilation operators, the interaction between these fields, written as the product of two fermion fields and the electromagnetic field, can be interpreted as a change of the fermion state: annihilation of the fermion in one state and its creation in another, accompanied by the emission or absorption of a photon. Physically, this can correspond to annihilation and creation of an electron, annihilation and creation of a positron, or the simultaneous annihilation (or creation) of an electron and a positron.

Quantum electrodynamical processes described by effects of first order in the fine structure constant were calculated during the late 1920s and early 1930s. The most important result was the Klein–Nishina–Tamm equation (1929) for the effective cross section of photon scattering by electrons, which agreed well with the data of D. V. Skobel'tsyn on the Compton effect. This was crucial for the acceptance of the theory based on the Dirac equation.

The Fermi model

Although the quantum field description of electromagnetic interactions had been formulated definitively by the beginning of the 1930s, the construction of a similar description for other processes was not at all obvious. In particular it was not understood how the concepts of "creation" and "annihilation" could be applicable to them. The theory of β decay proposed by Fermi in 1934 was of fundamental importance in this regard. Fermi used Pauli's suggested explanation for the continuous spectrum of electrons emitted in the decay and introduced the neutrino, which, escaping undetected, carries off part of the liberated energy. Fermi based his theory on the idea that the proton, electron, and antineutrino appearing in the decay do not actually enter into the composition of the neutron, but are created as a result of the interaction. In addition, in his β-decay model Fermi assumed that the entire process occurs at a single spacetime point, i.e., that the elementary interaction event is the annihilation of the neutron and creation of the proton, electron, and antineutrino.

Therefore, the weak interaction Lagrangian proposed by Fermi contains the product of four fermion field functions (expressed in terms of the creation and annihilation operators of the neutron, proton, electron, and neutrino) evaluated at a single point in spacetime.

The four-fermion interaction constant has dimensions $[G_F] = m^{-2}$. Its numerical value, determined from the β-decay probability, is approximately $10^{-5}\ \mathrm{GeV}^{-2}$.

In addition to the decay process $n \to p + e^- + \bar{\nu}_e$, the Fermi Lagrangian also describes other reactions, for example, $\nu_e + n \to p + e^-$ and $\bar{\nu}_e + p \to n + e^+$, as mentioned in Sec. 2.

The model later underwent further refinement. The structure of the four-fermion Lagrangian was made more accurate and modified so as to reflect the composite structure of hadrons. We shall discuss this in more detail in Secs. 6 and 17.

The Yukawa model

As far as the nuclear forces binding protons and neutrons in the nucleus are concerned, as already mentioned in Sec. 1, they differ from electromagnetic forces primarily in that they are short-range, the range being of the order of the nuclear size (10^{-13} cm).

Among the early strong interaction models, we should mention the "β-theory of nuclear forces" proposed by Tamm and D. D. Ivanenko, which was short-lived but historically important. In 1934 these authors suggested that in second-order perturbation theory the neutron and proton might interact, emitting and then reabsorbing virtual electron–neutrino pairs. However, this interaction proved to be too weak to explain nuclear forces.

In 1935, H. Yukawa used the idea of the exchange nature of short-range nuclear forces and proposed the first systematic model of a strong interaction mechanism. He suggested that the interaction between the constituents of the nucleus, protons and neutrons, is transmitted by a new, as yet unknown particle of spin 0 and mass m. Assuming that this particle is described by the Klein–Gordon–Fock equation and using this equation to derive the corresponding static potential:

$$(\Delta - m^2)U(r) = \delta(\mathbf{r}) \to U(r) = e^{-mr}/r,$$

Yukawa related the nuclear force range $r_{\text{nucl}} \approx 10^{-13}$ cm to the mass of the "interaction carrier" m. It turned out that this mass must be about 200 times greater than the electron mass. As is well known, particles with such properties, called π mesons, were discovered in 1947.

During the following decades the Yukawa model served as a qualitative tool for discussing the dynamics of strong interactions. However, it proved practically impossible to obtain satisfactory quantitative results using it. An exception was the semiphenomenological analysis of pion photoproduction on nucleons, carried out in 1950 by A.M. Baldin and V.V. Mikhaĭlov, who obtained a satisfactory quantitative description of this reaction near threshold.

5. Establishment of quantum electrodynamics

Ultraviolet divergences

Quantum electrodynamics encountered a serious difficulty. It was found that calculations using perturbation theory in orders higher than the first led to meaningless infinite expressions. This was first noticed in 1930 by Oppenheimer. Since these infinities first appeared when the contributions of high energy photons were taken into account, they were called ultraviolet divergences. These infinities are a consequence of the local nature of the interaction, and are a reflection and more general case of the familiar difficulties in describing point charges using classical electrodynamics. In the following chapter we shall give a more detailed discussion of the reasons for the appearance of ultraviolet divergences in a theory and the techniques for dealing with them. Here we restrict ourselves to the simplest examples and historical comments.

An example of a typical divergence is the electron self-energy. As is well known (see also Sec. 11), in the classical theory the energy of the electrostatic field produced by an electron of radius r_0 tends to infinity as $1/r_0$ for $r_0 \to 0$ (it is linearly divergent). In quantum electrodynamics the corresponding quantity diverges more weakly: it tends to infinity as $\ln r_0$ (this is a consequence of the inclusion of positrons in intermediate states).

Since in the calculations for a process ultraviolet divergences generally arise only in orders higher than the leading order, while the experimental accuracy was not very high in the 1930s, calculations were restricted to lowest order perturbation theory and the agreement with experiment was fairly close. The divergent momentum integrals encountered in higher orders were simply cut off at the upper limit. The typical expression which results involves the logarithm of the ratio of the energy at which the integral is cut off and the electron mass. Since the coefficient of this expression is small ($\alpha/\pi \sim 10^{-3}$), even for cutoff parameters much larger than the electron mass the relative corrections are small.

Nevertheless, the infinities present in higher orders presented a serious problem, since they made it impossible to judge the reliability of theoretical predictions.

At the end of the 1930s a number of authors, including Dirac, E. Stueckelberg, and H. Kramers, came up with several ideas for eliminating divergences. However, a general method for dealing with infinities was not successfully developed during those years.

The experimental results gave a new impetus to theoretical efforts in this area. In 1947 W. Lamb and R. Retherford measured the difference between the energies of the $2s_{1/2}$ and $2p_{1/2}$ states of the hydrogen atom. The Dirac equation applied to the hydrogen atom predicted that these two levels were exactly degenerate. It was clear that the level splitting arises from radiative corrections to the interaction between the electron and the Coulomb field of the nucleus. But when calculated, these corrections contained infinities, and it was necessary to learn how to eliminate them while extracting the physically meaningful part of the divergent expressions. This procedure, referred to as renormalization, was carried out successfully in the nonrelativistic approximation by Bethe in 1947. Since, as later became clear, most of the effect was due to low energy photons, Bethe's results agreed with experiment to within a few percent.

Covariant perturbation theory for the *S*-matrix

The need to construct a computational formalism in quantum electrodynamics which would involve relativistically invariant expressions at all stages became particularly acute in the 1940s. For this it proved convenient to use the scattering S-matrix introduced by Heisenberg in 1943. The S-matrix is the set of amplitudes for all possible transitions (caused by the interaction) of a quantum system from an initial state to a final state.

A relativistically covariant perturbation theory allowing the S-matrix to be constructed in the form of an expansion in powers of the small interaction constant was constructed in the studies of S. Tomonaga, J. Schwinger, and R. Feynman in the second half of the 1940s. In the approach of Tomonaga and Schwinger, the starting point was the Schrödinger equation for the state vectors, while Feynman began from the representation of the scattering matrix elements in the form of path integrals over field configurations. We shall postpone our discussion of path integrals until Sec. 14. Here we only note that this approach made it possible to develop a special graphical procedure for constructing the S-matrix in an arbitrary order of perturbation theory by means of Feynman diagrams. According to certain rules, also named after Feynman, each diagram is associated with a multiple integral over the virtual particle momenta (see Sec. 10). By now, Feynman diagrams have become one of the standard tools for calculating processes in quantum field models and are also widely used in other areas of physics.

The creation of renormalization theory

The successful realization of the idea of renormalization in the nonrelativistic approximation by Bethe and the development of covariant perturbation theory for the scattering matrix made it possible to construct a systematic formalism for subtracting ultraviolet divergences. This was done in studies by Feynman, Schwinger, F. Dyson, A. Salam, and others. In particular, Dyson and Salam performed a general analysis of the structure of divergences in arbitrarily high orders of perturbation theory. However, a number of subtle problems related to so-called overlapping divergences were not solved for a long time.

Here it proved necessary to carefully analyze the mathematical origin of divergences. This was accomplished by using a new branch of mathematics, the theory of generalized functions of S. L. Sobolev and L. Schwartz. It turned out that the Green functions for quantum fields (i.e., the Stueckelberg–Feynman propagators) are generalized functions like the Dirac delta function, and products of them are not, in general, defined. Ultraviolet divergences are a direct reflection of the ambiguities of such products for coincident arguments in the spacetime representation.

In the mid-1950s N.N. Bogolyubov and his collaborators used this as the basis for developing a technique of redefining products of causal propagators so as to ensure that the scattering matrix is finite and unique in arbitrary orders of perturbation theory. The corresponding statement is the content of the renormalization theorem of Bogolyubov and O. S. Parasiuk, and its prescriptive part—the so-called R-operation—serves as a practical basis for obtaining finite, unique results in perturbative quantum field calculations.

Successes of quantum electrodynamics

Therefore, the development of a systematic computational scheme in quantum electrodynamics made it possible to calculate any quantum electrodynamical effect to, in principle, an arbitrarily high degree of accuracy. The accuracy of actual calculations was now limited only by their complexity. (It should be noted that essentially the entire formalism exists unchanged at the present time.)

The successes of this theory were remarkable. In 1949 Schwinger calculated the first radiative correction to the electron magnetic moment—the so-called anomalous magnetic moment of the electron—with a relative accuracy of 10^{-3}. The Lamb shift was recalculated, more rigorously and considerably more simply, the radiative corrections to scattering amplitudes were obtained, and vacuum polarization effects were studied in detail. All the expressions obtained were in good agreement with experiment. We shall discuss the comparison of current quantum electrodynamical calculations with experiment in Sec. 16. Here we only note that the observed agreement is a record achievement in physics.

The ultraviolet asymptote

After it had become certain that various quantum electrodynamical effects could be described correctly using the lower orders of renormalized perturbation theory, attempts were made to obtain the possibly more complete information contained in the higher orders.

In some problems, for example, in the study of the asymptotic properties of the Green functions of quantum electrodynamics, the higher-order contributions are just as important as the lower-order ones. For example, in the ultraviolet region the actual expansion parameter is not the square of the charge α, but α times the logarithm of a large momentum k^2, i.e., $\alpha \ln k^2$. Therefore, the contributions of each order of perturbation theory are comparable. The correct asymptote is then obtained by summing all these contributions, i.e., terms of the type $(\alpha \ln k^2)^n$.

In the simplest cases it is possible to isolate these terms directly in arbitrary order. For example, in 1954 L.D. Landau, A.A. Abrikosov, and I.M. Khalatnikov calculated the ultraviolet asymptote of the propagators of quantum electrodynamics by direct summation of the leading logarithmic terms. On the basis of similar calculations, in 1955 E.S. Fradkin, and also Landau and I.Ya. Pomeranchuk, concluded that the photon propagator has an additional, unphysical "ghost" pole. The presence of such a singularity would indicate that quantum electrodynamics is inconsistent. However, more rigorous consideration shows that this conclusion cannot be strictly justified, since it requires information which cannot be extracted from perturbation theory.

The renormalization group in QED

It proved possible to settle this question by means of a systematic method of improving perturbation theory using the renormalization group. This group was first discovered in 1953 by Stueckelberg and A. Petermann, who pointed out the existence of continuous transformations effecting a reparametrization of renormalized perturbation theory and also the possibility of constructing group differential equations. Later, M. Gell-Mann and F. Low studied the structure of the ambiguities in the elimination of divergences by the cutoff method in QED and obtained a functional equation for the photon propagator (without regard to its group nature). They then used this equation to make a qualitative study of the possible ultraviolet asymptotes.

In a series of studies in 1955, Bogolyubov, D. V. Shirkov, and A. A. Logunov developed a systematic procedure for obtaining the functional equations for the Stueckelberg–Petermann renormalization group based on the finite renormalization transformations of Dyson. They also obtained the explicit form of the corresponding differential equations. They then solved the problem of reducing the results of perturbation theory to a renormalization-invariant form. The algorithm of the solution using the group differential equations is known as the renormalization group method. It allows the systematic improvement of the approximating features of perturbation theory and effectively reduces to the summation of infinite (renormalization-invariant) subsequences of ultraviolet logarithms. It was later used to discover the property of asymptotic freedom in quantum chromodynamics (see Sec. 13 below).

6. Problems of the 1950s

The CPT theorem and parity violation

In Sec. 3, as an example of symmetries which can occur in nature, we mentioned the symmetries under reflection of the three spatial axes (the *P* transformation), reflection of the time axis (the *T* transformation), and also interchange of particles and antiparticles (the *C* transformation). The presence or absence of symmetry under these transformations characterizes the fundamental properties of particle interactions.

In the early 1950s G. Lüders and Pauli proved the *CPT* theorem, according to which all interaction processes occurring in nature must be invariant under the three simultaneous *C*, *P*, and *T* transformations ($CPT = \mathbb{1}$). This means that if a certain process can occur, then the process in which the particles are replaced by the corresponding antiparticles, the spin projections change sign, and the initial and final states are reversed occurs with the same probability. It also follows from the *CPT* theorem that the masses and lifetimes of particles and antiparticles are equal, and that magnetic moments differ only by a sign.

The *CPT* theorem is based on very general principles incorporated in a theory, such as relativistic invariance, locality of the interaction, and the "normal" relation between the particle spins and statistics, given by the Pauli theorem. The predictions of the *CPT* theorem have been checked in many experiments, and not a single violation has yet been observed.

The situation with the *P*, *C*, and *T* transformations carried out separately is different. Each of these symmetries is present in the electromagnetic and strong interactions, but not in the weak interaction. In particular, *P* invariance is violated. Since invariance under the *P* transformation is associated with the conservation of a special quantum number, parity, it is said that the weak interaction does not conserve parity.

In 1956 T.D. Lee and C.N. Yang explained the existence of *K*-meson decays into systems of two and three π mesons in terms of parity violation. They also suggested experiments to check for parity violation in other weak interaction processes. Shortly afterwards, *P*-odd effects were discovered in all processes involving the weak interaction. It became clear that parity nonconservation is a fundamental feature of the weak interaction.

As pointed out by Landau, and also Lee and Yang, in 1957, mirror symmetry could exist in a certain sense if the so-called combined parity were conserved, i.e., if there were invariance under simultaneous *C* and *P* transformations. Here *CP* invariance would imply that for any process there must be an associated mirror-symmetric process, in which all the particles involved are replaced by antiparticles and all the characteristics of the two processes are identical.

However, in 1964, again in *K*-meson decays, it was found that *CP* invariance is also violated. (In contrast to spatial parity violation, up to now *CP* violation has been observed only in processes involving *K* mesons, and the nature of this phenomenon is not yet fully understood.) We note that the *CPT* theorem implies that such processes are also not invariant under time reversal.

Broken symmetries

Above we discussed symmetries which are present in all interactions except the weak interaction. Special symmetries which are present (approximately) only in strong interaction processes also play an important role in particle physics. Here we have in mind the so-called unitary symmetries of hadrons.

The idea first appeared in the early 1930s, when Heisenberg suggested that the neutron and proton be viewed as different charge states of a single particle, the nucleon. This was based on the fact that, although their electric charge is different, these two particles have similar masses and identical spins, and they behave similarly in strong interactions.

The fact that strong interactions are independent of the electric charge is referred to as isotopic invariance, and groups of particles with the same spin and parity, approximately equal masses, and identical behavior in strong interactions but different electric charge are called isotopic multiplets. For example, the proton and neutron form an isotopic doublet (p,n), and the π mesons form an isotopic triplet (π^+, π^0, π^-).

The formal mathematical analogy with ordinary particle spin is used to describe this important feature of strong interactions. Isotopic invariance is viewed as invariance under rotations in some abstract three-dimensional (isotopic) space, which form the group SU(2). Just as symmetry under ordinary spatial rotations leads to conservation of angular momentum, which in the absence of orbital motion coincides with the spin, isotopic invariance implies the conservation of a certain vector, called isospin, in isotopic space.

The concept of an internal symmetry as a factor determining how particles with similar properties are grouped into families also proved to be fruitful for classifying the large number of new hadrons discovered in the 1950s. At this time it was gradually becoming clear that the systematics of these particles required the use of not only the isospin I, but also a new quantum number called strangeness S. It was introduced in order to formalize a phenomenon observed in collisions of π mesons with nucleons: the associated production of K mesons and hyperons, unstable baryons with mass larger than the nucleon mass. If it is assumed that $S = 0$ for π mesons and nucleons and that this quantum number is conserved in strong and electromagnetic interactions, then only combinations of particles with zero total strangeness can be produced in such collisions.

By the beginning of the 1960s it had been established that the approximate symmetry group of hadrons could be extended to SU(3) if it was assumed that the quantum numbers determining the internal space in which the symmetry group acts are isotopic spin and strangeness. The isotopic symmetry group SU(2) is a subgroup of this group. (We note that the inclusion of additional quantum numbers discovered later leads to a further extension of the approximate symmetry group of strong interactions.)

Among the irreducible representations of the group SU(3) are representations of dimension 1, 8, and 10. In 1961 Y. Ne'eman and Gell-Mann used this fact to group the observed hadrons into multiplets of the appropriate dimension. For example, spin-1/2 baryons form an octet, and spin-3/2 baryons form a decuplet. Pseudoscalar mesons (spin 0) and also vector mesons (spin 1) form singlets and octets.

The scheme for classifying hadrons according to representations of the group SU(3) not only included all the hadrons known at that time, but also led to the prediction of new ones. For example, the existence of the Ω^- hyperon was predicted and its main features were described. Some time later this particle was discovered experimentally.

The SU(3) symmetry of hadrons is less accurate than the isotopic symmetry. This is manifested, in particular, in the fairly large (up to several tens of percent) difference between the masses of particles of the same multiplet. For example, the proton ($m_p \approx 983$ MeV) and the Ξ^- hyperon ($m_{\Xi^-} \approx 1321$ MeV) belong to the same octet.

The existence of an approximate hadron symmetry is an empirically established fact. In other words, at present we have no explanation for either the existence of the symmetry itself, or for its violation. Whereas one can attempt to attribute the violation of isotopic symmetry to the electromagnetic interaction, there is no such explanation for the higher symmetries. The quark model discussed in Sec. 8 also does not solve this problem, but only reduces it to the existence of a quark symmetry broken by the quark mass differences.

The V–A model of weak interactions

In the 1940s new weak interaction processes in addition to β decay became known. Examples are muon decay $\mu^{\mp} \rightarrow e^{\mp} + \bar{\nu}_e(\nu_e) + \nu_\mu(\bar{\nu}_\mu)$, μ capture $\mu^- + p \rightarrow n \rightarrow \nu_\mu$, and strange particle decays. This made it important to construct a theory of the weak interaction which would describe all these processes. It was found that a universal description of them could be obtained by introducing into the interaction Lagrangian terms analogous to the Fermi Lagrangian for β decay. Each of these terms has the form of a product of two expressions, called currents, which are bilinear in the fermion fields.

After the discovery of parity violation in weak interactions, a great effort was made to define the Lorentz structure of the weak currents more precisely. This resulted in the construction of the universal V–A model of weak interactions in 1957–58 by R. Marshak and E. Sudarshan, and also Feynman and Gell-Mann. In this theory each of the weak currents is the sum of two terms, one behaving like a vector (V) under Lorentz transformations and reflections, and the other behaving like an axial vector (A). Under all transformations except reflection of all the spatial axes, an axial vector behaves as a vector, while under such a spatial reflection it changes sign. Therefore, products of a vector and an axial vector lead directly to terms in the Lagrangian which describe processes with parity violation. In Sec. 17 we give a more detailed discussion of these questions and also the explicit form of the Lagrangian. Here we just recall an important feature of the vector current, namely, that it is conserved, as was first proved by S.S. Gershteĭn and Ya.B. Zel'dovich.

Problems with strong interactions

Turning now to strong interactions, we note that accelerator techniques underwent intensive development during the 1950s. The proton accelerators constructed at that time made it possible to study the interaction of nucleons and mesons at previously unforeseen energy scales. The amount of experimental data began to grow rapidly. It was during this period that the NN and πN interactions were studied quantitatively at energies up to 1 GeV and the resonances predicted by Pauli in 1946 were discovered. By the end of the 1950s it had become clear that resonances are a new, extremely important class of physical objects. Therefore, a theory of the strong interaction now had to give a quantitative description of not only a relatively small number of intranuclear forces, but also a wide range of phenomena in hadron physics, including the physics of resonances.

As already noted in Sec. 4, attempts to answer these questions using perturbation theory based on a Yukawa-type interaction, with one exception, did not meet with even qualitative success. Therefore, methods of analyzing problems in particle theory not directly related to smallness of the coupling constant and expansion in powers of it underwent significant development. The most important were the functional integral method and the dispersion relation technique. The first of these, borrowed from quantum mechanics, was initially used for purely theoretical applications (see Secs. 9 and 14 below). The latter not only gave integral relations between scattering characteristics which could be verified directly in experiment, but also led to an entire "bouquet" of approximation and semiphenomenological schemes, each of which played some role in systematizing the quantitative characteristics of hadronic processes.

Analytic methods

The dispersion relation technique, suggested in 1954 by Gell-Mann, M. Goldberger, and W. Thirring, essentially amounts to using the analytic properties of the scattering amplitude f viewed as a function of energy E. This approach requires generalization of the physical amplitude $f(E)$ to complex values of the argument, analysis of the properties of $f(z)$ in the entire complex z plane, and use of the theory of functions of a complex variable, in particular, Cauchy's theorem. In the end, one obtains the following integral equation for the real part of the complex scattering amplitude $f = \mathrm{Re}\, f + i\, \mathrm{Im}\, f$:

$$\mathrm{Re}\, f(E) = \frac{1}{\pi} \int_{-\infty}^{\infty} \frac{\mathrm{Im}\, f(E')}{E' - E}\, dE' + P(E). \tag{6.1}$$

This is a dispersion relation. The elaboration of the explicit form of the subthreshold integration region and the polynomial $P(E)$ requires the use of additional considerations, for example, the crossing symmetry conditions.

The most complicated part of the argument—the analytic continuation of the amplitude f into the complex plane—involves the use of the causality condition imposed on the scattering matrix. It is important to note that here general properties postulated for the scattering matrix as a whole are used instead of perturbation theory. This is referred to as the axiomatic approach to quantum field theory. The first proof of the dispersion relations was obtained by Bogolyubov in 1956 using the axiomatic approach, which is described in greater detail in Sec. 7 below.

It should be noted that the problem of analytic continuation is complicated from the purely mathematical point of view in that the scattering amplitude $f(E)$ for real, unphysical (for example, subthreshold) values of the argument E can possess singularities, i.e., it is a generalized function. The correct derivation of the dispersion relations therefore required the development of techniques for the analytic continuation of generalized functions.

The dispersion relations for forward πN scattering were derived rigorously in this manner using the hypothesis of a local pion field, and these relations survived comparison with experiment. This was of fundamental importance at the time, because it confirmed the correctness of the basic premises of local quantum field theory and also led to rapid growth of the popularity of analytic methods in the theory of strong interactions. However, the use of this approach to obtain rigorous results following directly from the general properties of quantum field theory, although tempting, proved to be quite complicated (see Sec. 7 below). Thus, there was widespread use of "easier" variants of the analytic approach, in which the desired analytic properties of the scattering amplitudes and the S-matrix elements in general were not proved, but postulated. The most ambitious construction of this type was the "analytic scattering matrix" program of G. Chew, in which analyticity was treated as a fundamental principle, replacing the basic premises of quantum field theory and even quantum mechanics. Here it was assumed that the use of nonlinear self-consistency conditions reflecting unitarity could lead to a unique description of strong interactions up to the determination of numerical parameters like coupling constants. A successful realization of this program providing an alternative to the usual formulation of quantum field theory has not been obtained.

Below, at the end of Sec. 7, we shall discuss some attempts to use the postulated analytical properties semiphenomenologically.

7. Axiomatic and analytic methods

The axiomatic approach

Methods of studying the analytic properties of the scattering matrix elements obtained a solid mathematical foundation within the framework of the axiomatic approach. This approach to the scattering matrix does not use the Schrödinger equation and its solutions. Rather, it is based on the general physical principles of the quantum theory of particle scattering, formulated as axioms.

An axiomatic technique of this type was first used by Bogolyubov in the early 1950s to explicitly construct the expansion of the scattering matrix in powers of the interaction (see Sec. 10 below for details). Developing this ideology further, in 1956 Bogolyubov, B. V. Medvedev, and M. K. Polivanov formulated a system of axioms without making any assumptions whatsoever about the strength of the interaction. This system starts from the Heisenberg representation of the S-matrix as an operator whose matrix elements are the probability amplitudes for transitions from one (the initial) state at $t = -\infty$ to another (the final) state at $t = +\infty$, each state containing both a set of individual particles at "infinite" separation from each other and clusters of these particles, i.e., bound states. Thus, the representation in which the interaction is switched on and off adiabatically is not used.

The Bogolyubov scheme contains a series of axioms about the structure of the set of quantum amplitudes (i.e., state vectors) describing the asymptotic sets of particles, their transformation properties, and also the properties of the S-matrix. The most important axioms are the conditions for relativistic invariance, unitarity, and causality, formulated directly for the scattering matrix.

This axiomatic approach allowed a large number of rigorous results of a general nature to be obtained. Examples are the one-dimensional spectral representations of single-particle Green functions ("dressed" propagators), four- and five-dimensional representations for the scattering amplitude (the so-called Jost–Lehmann–Dyson representations), and, finally, dispersion relations for the pion–nucleon scattering amplitude (see below) which relate only experimentally observable quantities.

In the following years axiomatic field theory underwent significant development. In the axiomatic approach of Wightman the fundamental physical object is the interacting quantum field, which is not directly observable. Conversely, the algebraic approach of R. Haag and H. Araki focuses on the set of all "observables," i.e., experimentally measurable physical quantities.

These axiomatic schemes were used successfully to obtain several general results for quantum field theory, for example, a new proof of the *CPT* theorem discussed in Sec. 6 above, and also theorems relating spin to statistics. Here we should also mention the Haag theorem, from which it follows that the interaction representation of a quantum field theory (i.e., the state vector of a system of quantum fields at a given finite time t) exists only when there is no interaction at all.

Dispersion relations

Dispersion relations are formally a consequence of Cauchy's theorem from complex analysis, applied to the scattering amplitude $f(E)$ analytically con-

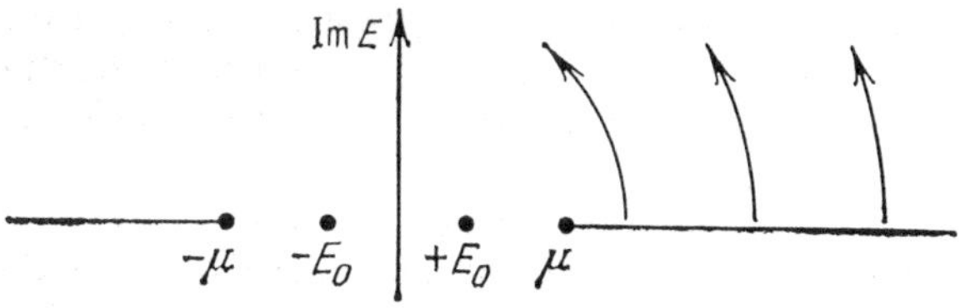

Figure 1. Structure of the unphysical region of integration in the dispersion relations for pion–nucleon forward scattering and scheme for analytic continuation from the physical region on the upper edge of the right-hand cut for $E > \mu$.

tinued into the complex energy plane $E \to z = E + iV$. Dispersion relations are usually written as integral relations of the type (6.1) between the real and imaginary parts of the two-particle scattering amplitude $f(E) = D(E) + iA(E)$. In the case of forward scattering, each of these can be expressed as algebraic relations:

$$D^2(E) + A^2(E) = |f|^2 = d\sigma_0/d\bar{\Omega}, \quad A(E) = \frac{\sqrt{E^2 - m^2}}{4\pi}\sigma_{\rm tot}(E)$$

involving two directly measurable quantities: the total effective scattering cross section $\sigma_{\rm tot}$ and the differential cross section for scattering at zero angle $d\sigma_0/d\bar{\Omega} = d\sigma(E,\theta = 0)/d\bar{\Omega}$.

The dispersion relations for pion–nucleon scattering were the first to be proved rigorously. For $\pi^{\pm} p$ forward scattering they have the form:

$$D_{\pm}(E) = P_{\pm}(E) + \frac{E^2 - \mu^2}{\pi}\int_{\mu}^{\infty}\frac{dE'}{E'^2 - \mu^2}\left\{\frac{A_{\pm}(E')}{E' - E} + \frac{A_{\mp}(E')}{E' + E}\right\}, \quad (7.1)$$

where $P_{\pm}$ is the sum of the pole and polynomial (in E) contributions:

$$P_{\pm}(E) = \frac{E + \mu}{2\mu}D_{\pm} + \frac{\mu - E}{2\mu}D_{\mp} + \frac{g^2}{2M}\frac{E^2 - \mu^2}{E \pm E_0},$$

containing three parameters: the threshold values of the amplitudes $D_{\pm} \equiv D_{\pm}(\mu)$ and the constant g characterizing the strength of the Yukawa-type pseudoscalar pion–nucleon interaction. The location E_0 of the "one-nucleon" pole is expressed in terms of the pion mass μ and nucleon mass M as $E_0 = \mu^2/2M$.

We see that, in addition to the physical region $E > \mu$, the range of integration includes the unphysical part of the continuum $E < -\mu$. The structure of the analytic continuation is shown in Fig. 1.

This analytic condition is possible because of the causality condition written directly for the scattering matrix. For studying the region of real negative energies, it is necessary to make use of the isotopic structure of the pion–nucleon amplitude, which relates $\pi^+ p$ and $\pi^- p$ scattering. We also note that, owing to the power-law growth of $f(E)$ for large E following from the optical theorem, the Cauchy theorem must be applied not to the actual scattering amplitude, but to the expression $f(E)/(E^2 - \mu^2)$, which falls off at infinity.

The integral over the large circle in the complex energy plane can thus be discarded and the remaining integration can run arbitrarily close to the real axis.

Since the threshold values $D_{\pm}(\mu)$, like the functions $D_{\pm}(E)$ and $A_{\pm}(E)$, are determined for $E>\mu$ directly from the experimentally observed cross sections, the dispersion relations (7.1) contain the single free parameter g^2. Therefore, verification of the dispersion relations amounts to showing that there exists some value of g^2 for which the left- and right-hand sides of the two equations in (7.1) are equal to each other.

Right after they were obtained, the dispersion relations were compared to the available data in the energy range accessible at that time. There was good agreement with experiment for $g^2 = 14.7$, which coincided with the estimates obtained earlier from pion photoproduction on nucleons near threshold. Therefore, the general principles of local quantum field theory had been verified directly in experiment.

After the dispersion relations for forward πN scattering had been proved rigorously, the dispersion relations were obtained for the πN-scattering amplitude at fixed square of the momentum transfer $t = (\Delta\mathbf{p})^2$ in a range $0 < t < t_{\max}$, for Compton scattering on nucleons, for pion photoproduction on nucleons, for a number of scattering processes involving virtual particles, and also for some inelastic processes of the type 2 particles $\to$ 3 particles.

The analytic approach

As already noted in Sec. 6, the success of the dispersion method led to a new perception of possible ways that particle interaction theory could be developed further, which was not, however, subsequently confirmed in full.

Meanwhile, it had become clear that the use of contemporary mathematical ideas and methods in quantum field theory was very important. The proof of the dispersion relations for nonzero momentum transfer required the simultaneous use and development of the theory of generalized functions and the theory of analytic functions of several variables. While doing this, Bogolyubov discovered a new principle for the analytic continuation of generalized functions into the complex region of several variables, and the "edge of the wedge" theorem was proved, leading to a fundamental new direction in mathematical physics.

At this time the concept of rigor in theoretical physics underwent a change: its mathematical standard was raised dramatically. The ideas which had been introduced into physics for proving dispersion relations provided the basis for a new language of quantum field theory. The new concept of the amplitude arose as an analytic function of several physical variables reflecting deep relations between seemingly completely different processes. Such relations between the amplitudes of processes connected to each other by crossing symmetry relations, realized by means of suitable analytic continuation, were the starting point for numerous heuristic schemes and constructions. The most important exact results obtained by this new analytic approach pertain to rigorous limits on the asymptotic behavior of scattering amplitudes at high energy. An example is the Froissart bound, according to which the total cross section cannot increase with energy faster than $(\ln E)^2$. Dispersion relations were used by Pomeranchuk to derive a theorem stating that the total cross sections for particle and antiparticle scattering on the same target are asymptotically equal when these cross sections approach constant values. A.A. Logunov, Nguyen Van Hieu, I.T. Todorov, and O.A. Khrustalev used the theory of analytic functions to generalize the Pomeranchuk theorem to the differential cross sections for particle and antiparticle scattering on identical targets.

The analytic approach to problems in the theory of strong interactions provided the basis for a number of semiphenomenological and approximation schemes developed in the 1950s and 1960s.

Applications of the analytic technique

Many of these schemes are somehow related to the double spectral representation for the two-particle scattering amplitude. This representation determines the analytic properties of the amplitude in two independent kinematical Lorentz-invariant arguments simultaneously. It was formulated by S. Mandelstam in 1958 on the basis of a correspondence to quantum field perturbation theory.

The validity of the double spectral representation was later established for the quantum mechanical scattering problem for a certain class of potentials of the Yukawa type. However, numerous attempts to prove it using quantum field theory were unsuccessful. Consequently, the Mandelstam representation was thereafter viewed either as a plausible hypothesis within the framework of quantum field theory, or as an independent postulate (in the analytic S-matrix scheme).

The unitarity conditions for the scattering amplitudes written down using this representation lead to nonlinear relations between their imaginary parts and the spectral densities. Such relations have been used to obtain various approximate equations, for example, a system of equations for the lowest partial waves in scattering at energies where inelastic channels are unimportant. Such formally closed systems for low energy scattering and transformations between pions and nucleons were studied intensively during the 1960s. They played a useful role in organizing the experimental data. However, it quite rapidly became clear that it was not possible to obtain physically closed descriptions of even a very restricted set of phenomena in this way.

Regge poles

Double spectral representations were an important component in the mathematical apparatus of a semiphenomenological scheme based on the concept of Regge poles and trajectories. This scheme was used to determine effective relations between the characteristics of low energy scattering in one channel, described by a 4-point amplitude, and the features of high energy scattering in another, the crossed channel. Here the guiding considerations, in addition to crossing symmetry and analyticity, were certain results obtained by T. Regge in the theory of nonrelativistic scattering on Yukawa-type potentials. In particular, he had demonstrated the fruitfulness of using the representation of continuous angular momentum l, obtained by the Sommerfield–Watson transformation in the variable cosine of the scattering angle. In this representation, bound and resonance states with identical values of the quantum numbers, except for the angular momentum, lie on straight lines in the plane of the variables E and l. Such curves are called Regge trajectories. The scattering amplitude in this representation is expressed as the sum of pole singularities of the form $1/[l - \alpha(E)]$, called Regge poles.

This beautiful and simple picture of the ordering of bound and resonance states was taken over to relativistic particle theory as a hypothesis by Chew and S. Frautschi, and also by V. N. Gribov. Supplemented by crossing symmetry considerations and the use of analytic continuation, it allowed the determination of qualitatively new relations between the parameters of resonances lying on a Regge trajectory in one, the direct, channel, and the characteristics of high energy diffractive scattering in another, the crossed, channel. Comparison with experiment showed that the angular momenta of hadron resonances with the other quantum numbers identical do actually tend to lie on smooth, monotonic curves which are nearly straight lines. Also, the data on diffractive scattering in the relevant crossed channels are, in a number of cases, in surprisingly good agreement with the corresponding expressions containing analytic continuations of the Regge trajectories.

In the following years many attempts were made to justify the Regge picture. However, it proved impossible to do this using local quantum field theory. Moreover, it soon became clear that in hadron scattering channels with the vacuum quantum numbers, where the so-called Pomeranchuk pole should dominate, the situation is very different from Regge pole theory.

Sum rules

Sum rules and, particularly, the so-called finite-energy sum rules, were an important semiphenomenological application of the analytic method.

Sum rules are closely related to dispersion relations and are a direct consequence of Cauchy's theorem applied to amplitudes which fall off sufficiently rapidly with increasing energy. They take the form of integrals over the energy variable of the imaginary parts of such amplitudes. When the falloff of the amplitude in question is insufficiently rapid, it is possible to approximate the high energy contribution by using a parametrization of the imaginary part based on the Regge pole model.

In this manner, by combining dispersion relations at fixed momentum transfer with the Regge asymptote hypothesis, in 1967 Logunov, L.D. Solov'ev, and A.N. Tavkhelidze, and also K. Igi and S. Matsuda, obtained finite-energy sum rules which explicitly contained the relation between the features of scattering in the direct and crossed channels. This relation later led to the concept of duality, which in turn provided the basis for the construction of dual models for hadron scattering amplitudes and also stimulated the development of string mechanisms leading to these dual amplitudes.

The quasipotential equation

One of the important rigorous results of the analytic method was an integral equation for the matrix elements of two-particle scattering. This equation, first obtained by Logunov and Tavkhelidze in 1963, is a direct generalization of the integral form of the Schrödinger equation for the nonrelativistic scattering amplitude. The analog of the potential involved in it depends on the energy and is called the quasipotential.

The iterative solution of this equation for a suitable choice of quasipotential exactly reproduces the perturbation series of quantum field theory for the relativistic scattering amplitude for a given interaction Lagrangian.

The Logunov–Tavkhelidze quasipotential equation is physically important in that it serves as the basis for various phenomenological approximations using some explicit form of the quasipotential. Such approximations are particularly fruitful in high energy diffractive scattering, where they successfully describe both the general behavior of the differential cross sections, and a number of subtle spin effects attributable to rescattering.

8. From quarks to quantum chromodynamics

Composite models

As already mentioned in Sec. 6, owing to the progress in accelerator techniques, during the 1950s a large number of new particles were found—strange hyperons and mesons, and also resonance states. There were a few score of them altogether. The description of each of these objects by means of its own quantum field led to a very complicated picture.

The first attempt to simplify this situation, by representing "elementary" particles or at least some of them as combinations (more precisely, bound states) of a smaller number of more fundamental objects, was the Fermi–Yang model of the late 1940s, in which it was assumed that π mesons are composed of nucleons and antinucleons.

In the studies of M.A. Markov (1955) and S. Sakata (1956), models were constructed and studied in which the strange hadrons known at that time were assumed to be fundamental, along with the proton and neutron.

The proton, neutron, and Λ hyperon were used in the Sakata scheme. Developing this scheme further, in 1957 L.B. Okun' assumed that a triplet of particles with the p, n, and Λ quantum numbers (but not these particles themselves) was fundamental, and then used this as the basis for constructing a nonet of pseudoscalar mesons, thereby predicting the existence of the η and η'. The Sakata–Okun' model also made it possible to determine the isotopic properties of weak hadronic currents, which are conserved in that theory. Later on, this model admitted SU(3) generalization for weak currents and meson resonances, difficulties in describing baryon resonances.

Because there was no quantum field theory of the strong interaction, it was not possible to carry out any dynamical calculations using these composite models. Their role was therefore restricted to the semiphenomenological organization of the spectroscopic characteristics of hadrons and resonance states, in a sense augmenting the regularities obtained from the Regge trajectory representation. It later became clear that the Sakata model was a prototype of the quark model.

SU(3) symmetry and quarks

In Sec. 6 we discussed the hadron systematics following from the approximate SU(3) symmetry. The observed hadron octets and decuplets correspond to representations of SU(3) of dimension 8 and 10.

The classification of hadrons according to representations of SU(3) led to some important consequences concerning the structure of hadrons. First of all, within the framework of SU(3) symmetry all hadrons are on an equal footing, so from this point of view it makes no sense to regard certain ones as being more fundamental. Furthermore, no one knew what the fundamental representation of SU(3), which has dimension 3, corresponded to.

An important assumption was made in 1964 by Gell-Mann and, independently, G. Zweig. Three new fundamental objects were introduced corresponding to "elementary carriers" of the quantum numbers—electric charge Q, baryon number B, strangeness S, and also spin J. Gell-Mann called these objects "quarks" and the name stuck. According to this new hypothesis, the three quarks u, d, and s realize the fundamental representation of the internal symmetry group SU(3), according to certain representations of which the observed hadrons are classified.

It was necessary to assign to the quarks fractional values of the electric charge and baryon number:

Quarks	Q	B	S	J
u	2/3	1/3	0	1/2
d	−1/3	1/3	0	1/2
s	−1/3	1/3	−1	1/2

Therefore, quarks q were introduced as very exotic entities, far removed from the observed particles. However, it turned out that, if it was assumed that only triplet combinations of quarks ($3q$) and antiquarks ($3\bar{q}$) and also paired combinations ($q\bar{q}$) are realized in nature, there results a very simple scheme for classifying the observed baryons, antibaryons, and mesons. This scheme is based on the following simple feature of unitary groups: any representation of a group SU(N) can be obtained by the tensor multiplication of fundamental representations.

Because the restrictions on the existence in nature of the quarks themselves and other combinations of them such as $2q$ were purely formal (and also because of difficulties related to violation of the Pauli theorem on the connection between spin and statistics; see below), in the beginning quarks served only as a device for the simple and elegant construction of all the hadrons with spin 0 and 1/2, and also the resonance states with spin 1 and 3/2.

It is easy to see how the observed hadrons are constructed from quarks. In the following table we give a few of them as an illustration:

Hadrons	Quark content	Relative spin orientations	Hadron spin	Q	B	S
p	uud	↑↑↓	1/2	1	1	0
n	udd	↑↑↓	1/2	0	1	0
π^+	$u\bar{d}$	↑↓	0	1	0	0
π^0	$u\bar{u}+d\bar{d}$	↑↓	0	0	0	0
ρ^0	$u\bar{u}+d\bar{d}$	↑↑	1	0	0	0
K	$d\bar{s}$	↑↓	0	0	0	1
K^+	$u\bar{s}$	↑↓	0	1	0	1
Δ^{++}	uuu	↑↑↑	3/2	2	1	0
Ω^-	sss	↑↑↑	3/2	-1	1	-3

The quark model provides a natural explanation for the existence of multiplets. In fact, states composed of a quark and an antiquark (not necessarily of the same type) can be constructed in nine different ways, giving a meson singlet and octet. Similarly, the 27 combinations of three quarks form a singlet, two octets, and a decuplet of baryons.

This splitting of the various combinations into multiplets has a simple justification in the theory of group representations. Specifically, products of three-dimensional representations of SU(3) can be decomposed into direct sums of representations of this group in the following manner: $3\times\bar{3}=1+8$; $3\times3\times3=1+8+8+10$.

All the hadrons known in the 1960s were successfully described by the three-quark model. In the early 1970s, to account for the suppression of certain weak decays it was suggested that there also exists a fourth quark, carrying "charm" (denoted by c). Right after this came the discovery of the J/ψ meson, which is interpreted as a $c\bar{c}$ bound state; later the charmed D and F mesons were found. At present, six types of quark are known, i.e., in addition to the u, d, and s quarks, there are three new quarks c, b, t with charges $Q_c = 2/3$, $Q_b = -1/3$, and $Q_t = 2/3$. Each quark has its own quantum number distinguishing it from the others. This quantum number, which actually numbers the type of quark, is referred to as the quark "flavor."

Color

The quark model in the form we have been discussing so far encounters a fundamental difficulty related to the existence of symmetric quark states. If quarks are assigned a spin of 1/2 (as is required to obtain the observed hadron spins), we immediately come into contradiction with the Pauli principle, since this model represents certain particles as combinations of three identical quarks located in the same state. Examples are the Δ^{++} resonance [represented as a state of three u quarks with parallel spins ($u\uparrow$, $u\uparrow$, $u\uparrow$)] and the Ω^- hyperon ($s\uparrow$, $s\uparrow$, $s\uparrow$). This would seem to violate the fundamental Pauli theorem relating spin to the type of statistics, which, as discussed above, follows from very general principles on which the theory is based.

The solution to this problem was found in 1965 by Bogolyubov, B.V. Struminskiĭ, and Tavkhelidze, and also by M. Han and Y. Nambu, who suggested that quarks carry some new quantum number (later called "color"). We note that the latter group of authors also came up with the idea of introducing an octet of vector gauge fields transmitting the interaction between quarks (i.e., the gluon fields; see Sec. 9 below). The color of each quark can take three values (by analogy with the three primary colors of the visible spectrum, they are called red, blue, and yellow). By definition, the colors of antiquarks are the complements of the quark colors. It is natural to call them "antired" (green), "antiyellow" (violet), and "antiblue" (orange).

The unusual feature of this new quantum number is that observable particles have no color degree of freedom. This makes it necessary to assume that for real hadrons (baryons and mesons) the color of their constituent quarks cancels, so that hadrons are colorless.

This is a good time to comment on the terminology used. This characteristic of quarks has, of course, nothing to do with real color, which is determined by the wavelength of emitted electromagnetic radiation. However, the choice of this terminology might be justified by the analogy with the theory of color vision, where the combination of equal parts of the three complementary colors gives white. This property can be represented intuitively as the result of the vector addition of colors (see Fig. 2).

In this new picture the quarks forming hadrons, for example, the Δ^{++} and Ω^-, differ in color, so there is now no conflict with the Pauli principle. Incidentally, we note that in this scheme it is possible for the quarks to have either fractional or integer charges, but the first variant is the most popular.

The unobservability of color implies the absence of a special direction in an abstract (color) space, i.e., invariance under rotations of this space. It is postulated that the color symmetry group is $SU_c(3)$, where the index c (for "color") is often written to avoid confusion with the approximate quark flavor symmetry. Here the quarks of three colors form the fundamental representa-

Figure 2. The vector addition of colors.

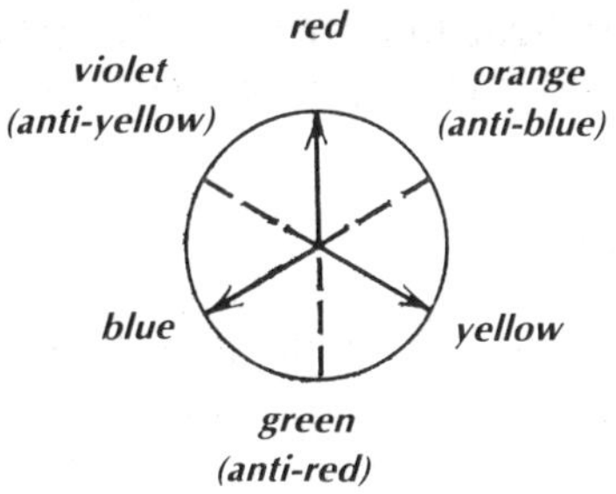

tion of the color symmetry group, and there are no colored objects which are more "elementary."

The hadronic wave functions, which are invariant under these transformations, are written as

$$|\text{meson}\rangle \sim \sum_{\text{over color } i} |q^i \bar{q}^i\rangle; \quad |\text{baryon}\rangle \sim \sum_{\substack{\text{over colors}\\ i,j,k}} \varepsilon_{ijk} |q^i q^j q^k\rangle,$$

where ε_{ijk} is the completely antisymmetric tensor.

In the language of representation theory, the fact that the observed hadrons are colorless implies that they belong to the singlet representation of the group $SU_c(3)$, which appears in the decomposition of the corresponding products of representations: $3\times\bar{3} = 1 + 8$, $3\times 3\times 3 = 1 + 8 + 8 + 10$. In this language the reason that colorless states cannot be formed from two, four, etc., quarks becomes obvious. Here the singlet representation is not contained in the decomposition of the corresponding products: $3\times 3 = 6 + 3$, etc.

Although color is not explicitly observed in nature, there are a number of experimental indications that quarks have three colors. We shall discuss them in Chap. 6, which is devoted to quantum chromodynamics. The requirement of local $SU_c(3)$ symmetry serves as the principle determining the interaction mechanism in this theory.

Partons

For the hadron systematics described above, it was sufficient to regard quarks as certain formal objects which can be used to construct states having the discrete characteristics of hadrons. After the problem with statistics was solved in the mid-1960s, Bogolyubov and co-workers proposed the first dynamical quark models. Subsequent studies and, especially, studies of processes with large momentum transfers in which the internal structure of hadrons is manifested, gave a real physical meaning to quarks.

As the energy of the colliding particles increases, the number of secondary particles created in a collision grows significantly because more and more new inelastic channels are opened. The diversity of possible transformation processes and the difficulty of observing a large number of reaction products greatly complicate the theoretical analysis of experimental data on hadron–hadron interactions.

Given these conditions, a new method of studying inelastic high energy interaction processes, proposed by Logunov in 1967, proved to be very fruitful. In this method a small number of particles of a given type in the final state are selected, and only these are detected experimentally. The probabilities for creating (any number of) all other particles are summed over. The cross sections obtained as a result of this summation are termed inclusive.

The utility of the inclusive approach was fully revealed in the analysis of experimental data on deep inelastic electron–proton scattering (where the energy and momentum transferred to the proton by the electron are much larger than the proton mass), obtained in 1968 at SLAC in California, and also data on high energy proton–proton scattering obtained in 1969 at Serpukhov in the USSR. The property of self-similarity (also called "scaling") of the inclusive cross sections was discovered, indicating the absence of a characteristic dimension or mass scale in hadron physics in the region of several GeV.

The experimental discovery of scaling motivated the development of the parton model of hadrons by Feynman in 1969. According to this model, a nucleon participating in high energy deep inelastic processes behaves as an object consisting of discrete scattering centers—partons, which with good accuracy can be considered pointlike, i.e., of zero dimension.

This picture is reminiscent of the ideas of Markov, who in 1964 showed that when inelastic channels play the dominant role, the total cross section for high energy lepton–hadron scattering is similar to the cross section for scattering on a point particle and does not depend on the effective size of the nucleon.

In the early 1970s, Bogolyubov, V.S. Vladimirov, and Tavkhelidze proved that the scaling behavior of the amplitudes is consistent with the general principles of local quantum field theory.

The parton picture of hadronic structure was later confirmed in the deep inelastic scattering of neutrinos and antineutrinos on nucleons. It also proved possible to obtain some information on the parton quantum numbers. It was found that charged partons have spin 1/2. This led to the assumption that charged partons can be identified with quarks.

An important argument for this identification was the validity of the "quark-counting" rules proposed in 1973 by V.A. Matveev, R.M. Muradyan, and Tavkhelidze. The quark counting rules, which are obtained from scaling considerations and representations of the composite structure of hadrons, determine the energy dependence of the transverse differential cross section for an arbitrary binary reaction involving hadrons in high energy scattering at large angles, and also the asymptote of the form factors at large values of the momentum transfer.

On the whole, the quark–parton model of hadrons, which explained a number of fundamental observed regularities, was an important stage in the development of the quantum field theory of the strong interaction—quantum chromodynamics, which will be discussed below in Chap. 6.

9. Gauge fields and interactions

Yang–Mills fields

In this section we shall discuss the history of the development of gauge field theory. This is an important direction in quantum field theory which in recent years has led to great advances in the understanding of the nature of particle interactions. The theory is based on the principle of local gauge symmetry, which corresponds to the principle of relativity in charge space. It was first used by Fock in 1926, when he succeeded in showing that the Klein–Gordon–Fock equation for a particle in an electromagnetic field is invariant under the simultaneous phase transformation of the ψ function and gauge transformation of the electromagnetic field potential:

$$\psi(x) \to \psi(x) e^{-ief(x)}, \quad A_\nu(x) \to A_\nu(x) + \partial_\nu f(x).$$

In 1954 Yang and R. Mills generalized the principle of local gauge symmetry to the SU(2) group of isotopic transformations. They showed that invariance under local phase transformations having an isotopic structure can be obtained by replacing the ordinary derivatives of fields in the free Lagrangian by so-called covariant derivatives containing a new vector field B_μ. (This procedure is referred to as "extension of the derivative.") This gives rise to an interaction between the original fields, sometimes called matter fields, and the field B_μ. This field, which is the generalization of the electromagnetic field to the case of the non-Abelian group SU(2), transforms under the adjoint representation of this group. In contrast to Maxwell's equations, the equations of motion of the field B_μ are nonlinear, which physically corresponds to the interaction of this field with itself.

The Yang–Mills construction can be generalized to more complicated transformation groups. The vector fields introduced here are called gauge fields or Yang–Mills fields. The set of local phase transformations of the matter fields and the corresponding transformations of the vector gauge fields are called gauge transformations.

As when the electromagnetic interaction is introduced, the extension of the derivatives in the case of non-Abelian local symmetry groups is accompanied by the appearance of, in addition to the vector field, a certain constant characterizing the interaction of all the matter fields with the gauge field.

A set of matter fields is associated with each gauge group. These fields transform like certain representations of this group and interact with the gauge vector field obtained by suitable extension of the derivatives. They also interact with each other via the gauge vector field. One thus obtains a model for the interaction of a set of fields which contains only one coupling constant. This attractive picture is complicated by a number of factors.

First, non-Abelian gauge fields, like the electromagnetic field, are massless, since a mass term spoils the gauge invariance. This severely restricts the direct physical application of such a theory.

Second, Yang–Mills fields, like the electromagnetic field, contain unphysical degrees of freedom, so constraints are present in the theory. Moreover, the self-interaction mixes up the physical and unphysical degrees of freedom. All of this greatly complicates the quantization procedure.

Quantization

Feynman was the first to study the details of the quantization of non-Abelian gauge fields in 1963. He showed that if the ordinary propagator of a massless vector field is associated with a Yang–Mills field, the S-matrix constructed using this propagator is not unitary. A systematic quantization scheme was constructed in 1967 by L.D. Faddeev and V.N. Popov, and independently by B. De Witt. An approach which is particularly convenient for quantizing gauge fields is that in which the quantum dynamics is described by a path integral—a special type of continuous sum over all field configurations. We shall become acquainted with the path integral concept in Sec. 14. Here we only note that this object possesses the algebraic properties of an ordinary integral and admits both integration by parts and change of variables. The latter feature is especially important for the gauge field quantization procedure, where to satisfy the constraints it is necessary to integrate not over all field configurations, but only over gauge-equivalent classes of fields.

Massive vector fields and the Higgs effect

As mentioned above, the fact that the gauge fields are massless made the physical interpretation of this theory difficult, since no massless particles of spin 1, except for the photon, were known.

An interesting possibility arises in models in which the gauge invariance is broken spontaneously. In the early 1960s the phenomenon of spontaneous symmetry breaking in quantum field theory was studied intensively by Nambu, V.G. Vaks and A.I. Larkin, J. Goldstone, and G. Jona-Lasinio, who used Bogolyubov's idea about spontaneous symmetry breaking in quantum statistics. In 1964 P. Higgs and several other authors showed that the gauge vector field can obtain a mass in models with spontaneous breakdown of the local symmetry. This is known as the Higgs mechanism and is the analog of the well-known Landau–Ginzburg effect in the phenomenological theory of superconductivity, which leads to the appearance of a gap in the electron energy spectrum.

In 1971 G. 't Hooft showed that the general methods for quantizing massless Yang–Mills fields can be applied, practically without change, to the case of spontaneously broken symmetry.

Electroweak interactions

After the discovery of the mechanism by which the gauge vector bosons acquire a mass via spontaneous symmetry breaking, it became possible to construct a gauge theory of the weak interaction. Among the proposed models, the one which was subsequently confirmed was that created in 1967 by

S. Weinberg and independently by A. Salam. It is a unified theory of the electromagnetic and weak interactions in which the electromagnetic field and the field of the intermediate vector boson transmitting the weak interaction are combined into a multiplet of Yang–Mills fields corresponding to the group SU(2)×U(1). The difference between the properties of these fields arises from spontaneous symmetry breaking.

Renormalization

In the early 1970s A.A. Slavnov developed methods for the invariant regularization and renormalization of gauge field theory. These methods were later extended to the situation where the gauge fields acquire mass via spontaneous symmetry breaking. 't Hooft and M. Veltman developed a new regularization technique, called dimensional regularization, which turned out to be very convenient for calculating renormalized diagrams. This essentially completed the construction of the perturbation theory for the finite and unitary scattering matrix in theories containing gauge fields.

Asymptotic freedom and quantum chromodynamics

Non-Abelian gauge theories possess an important property called asymptotic freedom. This was discovered in 1973 by D. Gross and F. Wilczek and, independently, by H. Politzer. This is the property that the strength of the interaction transmitted by non-Abelian gauge fields tends to zero as the momentum transferred in the interaction tends to infinity. It is this feature of the interaction between quarks (partons) which is most consistent with experimental observations.

After this discovery, in the mid-1970s the modern quantum field model of interquark interactions, "quantum chromodynamics," was developed. It is based on the assumption that a local color $SU_c(3)$ symmetry exists in nature. The quanta of the corresponding gauge field which transmit the interaction between quarks are called gluons. In the interpretation of quantum chromodynamics from the viewpoint of the parton picture, the role of the gluons is played by electrically neutral partons, which, as indicated by the experimental data, carry about half of the hadron momentum. Hadron interactions are now represented as a complicated interaction process involving their constituent quarks and gluons.

At the present time, quantum chromodynamics is consistent with all the known experimental data.

The structure and properties of the Weinberg–Salam model and quantum chromodynamics, and also their experimental status are discussed in detail below in Chaps. 5 and 6.

Gauge interactions

We thus see that the electromagnetic, weak, and strong interactions are successfully described by quantum field models of the gauge type. In 1956 R. Utiyama showed that the gravitational interaction can be introduced in a similar manner. For this it is necessary to extend the concept of gauge invariance and consider a local version of the Poincaré group, i.e., the transformation group obtained by replacing the parameters of the original group by a set of arbitrary functions. In this manner one can derive the equations of general relativity, with the gravitational field appearing as a gauge field.

Therefore, in the modern theory the heuristic description of the interactions of particles and the corresponding fields is replaced by a single principle for constructing the dynamics of fundamental interactions: every interaction between fields is based on some symmetry. This symmetry, when made local, leads to the necessity of introducing the corresponding gauge field responsible for the given interaction mechanism. In this formulation the question of the nature of the interactions is replaced by the question of the nature of the symmetries.

We should note that it has not yet been possible to realize this principle in a "pure" form. Both in the electroweak interaction model and in theories of new hypothetical interactions, which we shall discuss in the final chapter, it is not possible to describe all the details of the dynamics using the gauge principle. Interactions of the Yukawa type and the fourfold self-interaction of scalar fields play only an auxiliary role in these models, but they are quite important, since they lead, for example, to the existence of masses for the gauge field quanta.

Chapter 3
From renormalization to the renormalization group

As mentioned in the preceding chapter, the 1940s and '50s were marked by great progress in the development of quantum field theoretic techniques. The most important step was the solution of the problem of eliminating ultraviolet divergences. This was preceded by the construction of covariant perturbation theory and the idea of renormalization realized on the basis of perturbation theory. The final event of this period was the discovery of the renormalization group and the development of a powerful computational tool—the renormalization group method, the full value of which was realized only in the 1970s and '80s. Because of the special importance of these questions, which lie at the basis of contemporary local quantum field theory, we discuss them in more detail in the present chapter.

10. Covariant perturbation theory

Formulation of the problem

In formulating the problem of the interaction in quantum field theory, as in quantum mechanics one usually starts from the Schrödinger equation:

$$i\,\frac{\partial\psi(t)}{\partial t}=H\psi(t), \tag{10.1}$$

in which the full Hamiltonian of a system of fields H is the sum of the Hamiltonian of the free motion H_0 and the interaction Hamiltonian H_1. Since it is generally impossible to obtain the exact solution of this equation, one resorts to perturbation theory based on the initial approximation of free, noninteracting fields. Here the interaction must be viewed as a factor which slightly changes the properties of the physical system—a factor which can be switched on and off. At first glance this formulation of the problem seems permissible.

However, even free particles interact with the fields they create. As is well known, in the quantum case particles interact with the vacuum as though it were a type of physical medium. In spite of the possible weakness of the interaction due to the small particle size, interaction effects prove to be large, and in the limit of point particles they are infinite.

Therefore, the use of the concept of "bare" particles, i.e., particles which do not interact even with the vacuum, is a weak spot in perturbation theory. It is desirable to deal from the very beginning with real interacting particles and the corresponding quantum fields. This program is to a certain extent realized in the so-called axiomatic construction of quantum field theory (see Sec. 7 and also the end of the present section), and also in the functional integral representation (see Secs. 14 and 20).

However, in the standard form of quantum field perturbation theory, as typified by the Feynman diagram, it is necessary to be able to convert from "bare" to "physical" particles. This procedure is complicated by the need to eliminate the ultraviolet divergences which arise. This is accomplished by a special procedure known as the renormalization of infinities.

The scattering matrix

As discussed above in Sec. 5, the covariant form of quantum field perturbation theory is based on the use of the scattering matrix.

Let us consider a process at whose beginning and end there are only particles at finite separation from each other which can be assumed free. We denote the wave function of the initial state by $\psi(-\infty)$ and that of the final state as $\psi(\infty)$. They are related as

$$\psi(\infty) = S\psi(-\infty).$$

The quantity S introduced in this manner is the scattering matrix (or operator). The squares of its matrix elements determine the transition probabilities and effective cross sections for the possible scattering processes and particle transformations.

In order to determine the scattering matrix from the Schrödinger equation, this equation is first transformed to a different representation—the interaction representation, in which the time dependence related to the dynamics of the interaction is concentrated in the wave function, and the Hamiltonian reduces to an interaction Hamiltonian depending on operator functions which satisfy homogeneous free field equations.

The new equation is solved by the method of successive approximations. The limit $t \to \infty$ is taken in the resulting wave function $\psi(t)$, and the S-matrix is obtained as a series $S = 1 + \Sigma_{n\geqslant 1} S_n$, where the nth term S_n is proportional to the corresponding power of the interaction Hamiltonian. Using the fact that, for models which do not contain derivatives of functions in the interaction Lagrangian, the Hamiltonian differs from the Lagrangian by only a sign:

$$H(t) = -L(t) = -\int \mathscr{L}(x)d\bar{x},$$

S_n can also be written in terms of the interaction Lagrangian as

$$S_n = \frac{i^n}{n!}\int dx_1 \ldots \int dx_n\, T\{\mathscr{L}(x_1)\ldots\mathscr{L}(x_n)\},$$

where $\int dx$ stands for integration over all of spacetime and T stands for time-ordering. The latter means that the field operators entering into the product are ordered such that the time arguments x_k^0 decrease. For example, $T\{A(x)B(x)\} = A(x)B(y)$ if $x^0 > y^0$, and $T\{A(x)B(y)\} = B(y)A(x)$ if $y^0 > x^0$.

Therefore, the noncovariant quantity—the wave function $\psi(t)$ characterizing the system at a given time t—is not involved at all in the S-matrix formalism, while the scattering matrix, defined as the evolution operator of the system over an infinitely long time interval, can be expressed in a covariant manner in terms of the interaction Lagrangian $\mathscr{L}$. The time-ordered products of Lagrangians reduce to time-ordered products of the corresponding fields, and the S-matrix elements are expressed in terms of the vacuum expectation values of time-ordered products of operator-valued field functions. This provides the basis for the introduction of simple rules for constructing the matrix elements. These rules are related to the Feynman diagrams.

The Feynman rules

The Feynman diagrams in the coordinate–time representation seemingly depict the motion of particles between interaction events. Interaction events are denoted by points (the vertices of the diagrams), and particle propagation is denoted by lines. According to special rules (the Feynman rules), each element of a diagram is associated with a factor in the corresponding term of the scattering matrix. The essential feature is the fact that the functions corresponding to particle motions are the Green functions of the equations of motion for the corresponding fields with special boundary conditions. These are the so-called Stueckelberg–Feynman propagators, which can be determined as the vacuum expectation values of the time-ordered products of two field operators. They are also referred to as causal Green functions. They simultaneously describe the motion of particles and their antiparticles in opposite directions. For example, the propagator of the electron–positron field:

$$S^c(x-y) = i\langle T\psi(x)\bar{\psi}(y)\rangle_0,$$

satisfying the inhomogeneous Dirac equation:

$$(i\gamma^\nu\partial_\nu - m)S^c(x-y) = -\delta(x-y), \quad \partial_\nu \equiv \frac{\partial}{\partial x^\nu},$$

can be written as a Fourier integral:

$$S^c(x) = \frac{1}{(2\pi)^4}\int \frac{m+\hat{p}}{m^2-p^2-i\varepsilon}\, e^{-ipx}\, dp, \quad \hat{p} \equiv p^\nu\gamma_\nu$$

(in the denominator the infinitesimal quantity $\sim\varepsilon\to+0$ reflects the boundary conditions mentioned above). For $x^0 < y^0$ it corresponds to the motion of an electron from the point x to the point y, and for $x^0 > y^0$ it corresponds to the motion of a positron from y to x.

As an illustration, let us consider Compton scattering $\gamma_i + e_i \to \gamma_f + e_f$. In lowest (second)-order perturbation theory it is depicted by the two diagrams in Figs. 3 and 4. Diagram 4 simultaneously takes into account two possibilities. For $x^0 < y^0$ the initial electron e_i first emits the final photon γ_f at the point x and is transformed into a virtual electron which moves from x to y.

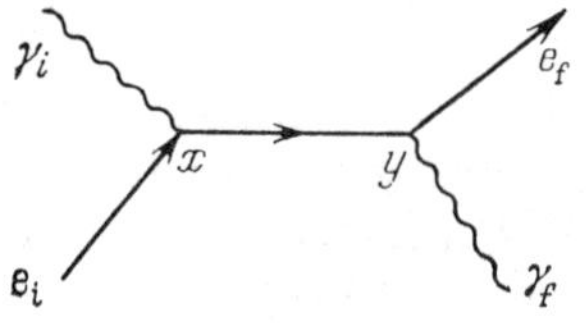

Figure 3. One of the two diagrams for the Compton scattering of a photon on an electron in second order perturbation theory.

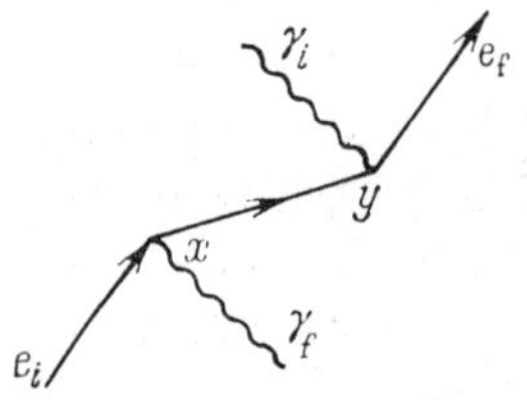

Figure 4. The other second-order diagram for the scattering of a photon on an electron.

Then at the point y the virtual electron absorbs the initial photon γ_i and is transformed into the final electron e_f. If instead $x^0 > y^0$, the order and nature of the events change (see Fig. 5). First, at the point y the initial photon γ_i is transformed into a pair consisting of the final electron e_f and a virtual positron, which then undergoes annihilation at the point x with the initial electron e_i, becoming the final photon γ_f.

For studying the matrix elements it is more convenient to use the Feynman diagrams and rules in the momentum representation. Examples of such diagrams are given in the next section. The crucial feature of the rules in the momentum representation is the fact that each topologically closed object, a loop, is associated with an integral over the appropriate 4-momentum; see the diagram in Fig. 6 below. The covariant perturbation theory for the scattering matrix using Feynman diagrams greatly simplified perturbative calculations, made it easier to carry out the renormalization program, and eventually went beyond the framework of quantum field theory.

The Bogolyubov axiomatic scheme

The analysis of the divergence structure and formulation of the technique for eliminating divergences proved to be simpler using a special scheme worked out by Bogolyubov in the early 1950s. In it the S-matrix is introduced axiomatically, without recourse to the Hamiltonian and the Schrödinger equation.

In this approach the S-matrix is determined from the physical requirements of covariance, unitarity, and causality which it satisfies. The causality requirement plays a special role. The intuitive meaning of this requirement is quite obvious—the evolution of any system can only be affected by events which occur at earlier times. However, the formulation of this principle in local quantum field theory requires the determination of nontrivial space-time relations.

Figure 5. Labeling of the diagram in Fig. 4 in the case where there is a positron in the intermediate state.

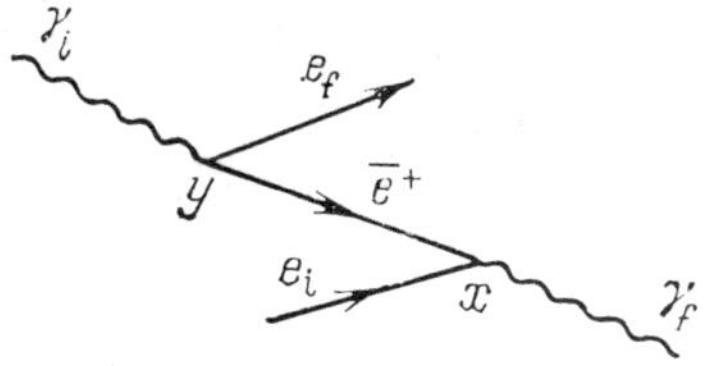

To formulate the causality condition we introduce the generalized scattering matrix $S(g)$ corresponding to the formal "switching on" of the interaction in the spacetime region characterized by the function $g(x)$. The matrix $S(g)$ is a functional of $g(x)$, and the Bogolyubov causality condition is formulated in terms of functional derivatives of $S(g)$ with respect to $g(x)$. The physical scattering matrix corresponds to the case where $g(x) = 1$ in the entire infinite spacetime continuum. Historically, this construction of the scattering matrix was the first axiomatic construction in quantum field theory. It later played an important role in proving the dispersion relations.

11. Divergences and renormalizations

Ultraviolet divergences

A difficulty peculiar to quantum field theory is the presence of ultraviolet divergences. In calculations using quantum mechanical perturbation theory it is necessary to sum over all possible intermediate states of the physical system under consideration. In quantum field theory these states contain additional virtual particles compared to the initial and final states. The corresponding sums are converted to integrals over all values of the energies and momenta $k = (k_0, \mathbf{k})$ of the intermediate states of the system of particles.

Let us turn to the process of the Compton scattering of a photon on an electron. Its matrix element in lowest (second)-order perturbation theory contains the contribution of the diagram shown in Fig. 3. Assuming that the time arrow points from left to right, we see that after the first interaction event at the point x the initial photon γ_i and electron $\bar{e}_i^-$ are transformed into a single particle—a virtual electron $\bar{e}^-$, which at the point y decays into a final photon γ_f and electron e_f^-.

The sum of the second-order contributions, which are proportional to the square of the electric charge, i.e., the first power of the fine structure constant $\alpha = e^2/\hbar c$, gives the matrix element αM_2, whose square leads to the familiar Klein–Nishina–Tamm equation.

We note that the 4-momentum of the virtual electron $\bar{e}^-$ is fully determined by the kinematics of the process, so the matrix element αM_2 does not involve any integrations. However, in the next highest nonzero order, the fourth order, of perturbation theory the situation is quite different. The ma-

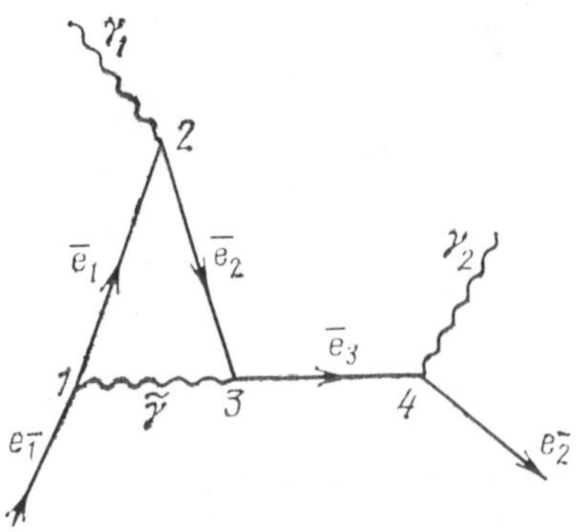

Figure 6. One of the diagrams for the Compton scattering of a photon on an electron in fourth-order perturbation theory.

trix element $\alpha^2 M_4$ contains a contribution from the diagram in Fig. 6. The sequence of elementary interactions in Fig. 6:

$$(1)\, e_{\bar{1}} \to \bar{e}_1 + \bar{\gamma}, \quad (2)\, \gamma_1 + \bar{e}_1 \to \bar{e}_2,$$
$$(3)\, \bar{\gamma} + \bar{e}_2 \to \bar{e}_3, \quad (4)\, \bar{e}_3 \to \gamma_2 + e_{\bar{2}}$$

is such that the intermediate state arising after the first interaction contains a virtual photon whose 4-momentum k is not completely defined. This photon $\bar{\gamma}$ is also present in the next intermediate state between interactions (2) and (3). It is necessary to sum over all possible values of the components k_ν. Therefore, the matrix element M_4 is expressed in terms of an integral over the 4-momentum k. But this integral is divergent! It turned out to be impossible to calculate the radiative correction, which is proportional to $\alpha^3(M_2 M_4 + M_4 M_2)$, to the Klein–Nishina–Tamm equation.

The divergence of such integrals entering into the radiative corrections of the matrix elements for various processes at large values of the virtual momenta is not an exception, but the rule. The formal reason for this is the insufficiently rapid falloff of the integrands for $|k_\nu| \to \infty$. For example, the closed loop with three vertices on the left-hand side of the diagram in Fig. 6 in the momentum representation corresponds to the integral:

$$\Gamma_\mu(p,q) = \int \frac{d^4k}{k^2} \frac{m + \hat{p} - \hat{k}}{m^2 - (p-k)^2} \gamma_\mu \frac{m + \hat{p} + \hat{q} - \hat{k}}{m^2 - (p+q-k)^2};$$

$$\hat{k} \equiv \sum \gamma^\nu k_\nu, \qquad (11.1)$$

where p and q are the momenta of the initial electron and photon. We see that the integrand contains the momentum squared in the numerator and to the sixth power in the denominator. Thus, the integral diverges at the upper limit, i.e., for large values of the momentum k.

In quantum electrodynamics the region where these integrals diverge corresponds to large, ultraviolet values of the virtual photon frequency $\omega = k_0 c/\hbar$. On the other hand, the 4-momentum and spatial coordinate representations are Fourier transforms of each other. Thus, according to the familiar properties of this transform, large momenta $|\mathbf{k}|$ correspond to small distances $r \sim \hbar/|\mathbf{k}|$ in ordinary space. Therefore, ultraviolet divergences are a consequence of the features of the interaction mechanism at small spacetime separations, or, more precisely, the local nature of the interaction between quantum wave fields.

In relation to this, it is useful to note that a well known difficulty arises in classical electrodynamics owing to the pointlike nature of the electron. As mentioned in Sec. 5, the energy of the electromagnetic field due to a point charge is infinite. A field correction to the electron mass is associated with the field energy according to the relation $E = mc^2$. The ultraviolet divergences in quantum field theory are the relativistic quantum analog of this infinite field mass of the classical point electron.

Regularization

Henceforth, when discussing divergent integrals we shall characterize the degree of divergence by the dependence on a fictitious upper limit Λ, corresponding to the ultraviolet cutoff of the 4-dimensional integration region by a condition of the form $|k_\nu| \leqslant \Lambda$. The conventional auxiliary representation of divergent integrals like Γ_μ by convergent expressions containing additional parameters is called regularization. Regularization is convenient and important because it allows us to deal with finite expressions during the intermediate stages of a calculation. Of course, to return to the actual situation at the end of the calculation it is necessary to remove the regularization. In the case of cutoff regularization the regularization parameter Λ is taken to infinity.

We note that other types of regularization are also used in quantum field theory: Pauli–Villars regularization, analytic regularization, and also 't Hooft–Veltman dimensional regularization. The last of these consists of replacing the integral over a 4-dimensional manifold by a symbolic operation which formally corresponds to integration in a momentum space of noninteger dimension $D = 4 - 2\varepsilon$ less than four. Here the regularization parameter is the infinitesimal quantity ε, and the regularization is removed by letting ε go to zero. In recent years dimensional regularization has been used widely owing to its advantages in theories possessing symmetries of the gauge type and also owing to its technical simplicity.

Renormalizations in quantum electrodynamics

In quantum electrodynamics it turns out to be possible to isolate and group together all the divergent quantities into a small number of structures which, in the end, can be interpreted as field additions δm and δe to the mass and charge of the electron. For example, the divergent integral $\Gamma_\mu(p,q)$ in Eq. (11.1) can be written as the sum of a finite function $\widetilde{\Gamma}_\mu(p,q)$ of the initial particle momenta and a divergent constant C multiplied by the Dirac matrix γ_μ. The part of the matrix element $\alpha^2 M_4$ containing this constant, $\alpha^2 C\gamma_\mu$, has the same functional and matrix structure as the lowest-order matrix element

αM_2. It can therefore formally be "moved" from the fourth to the second order. Then the resulting sum:

$$\alpha M_2 + \alpha^2 C M_2 = (\alpha + \delta\alpha) M_2, \quad \delta\alpha = \alpha^2 C,$$

can be represented as a shift of the electron charge, since

$$\hbar c(\alpha + \delta\alpha) \approx (e + \delta e)^2,$$

where $2e\delta e = \hbar c \cdot \delta\alpha$.

It is important that in quantum electrodynamics *all* the divergent contributions can be written as additions to the mass and charge of the electron. Here the matrix elements turn out to be functions of only the combinations $m_0 + \delta m$ and $\alpha_0 + \delta\alpha$, where m_0 and α_0 are the original, "bare" mass and charge of the electron. (Here and below we shall use the term "electron charge" to refer to the fine structure constant, which is actually proportional to e^2.) The bare parameters of the electron and the ultraviolet divergences enter the matrix elements only in these combinations. It therefore becomes possible to identify the sums:

$$m_0 + \delta m = m_{\text{phys}}, \quad \alpha_0 + \delta\alpha = \alpha_{\text{phys}}$$

as the experimental values of the mass m and the constant $\alpha = 1/137$.

It is this redefinition which provides the basis for the operation of eliminating ultraviolet divergences, also referred to as renormalization. This term is associated with the representation of the effect of radiative corrections not by sums, but by products:

$$m_0 + \delta m = Z_m m_0, \quad \alpha_0 + \delta\alpha = Z_\alpha \alpha_0,$$

where Z_m and Z_α are certain dimensionless singular factors. The operation of redefining the electron parameters by multiplying by numerical factors:

$$m_0 \to Z_m m_0 = m_{\text{phys}}, \quad \alpha_0 \to Z_\alpha \alpha_0 = \alpha_{\text{phys}},$$

is also called renormalization (or sometimes "renormalization of infinities").

The change of the mass and charge of a particle under the influence of an interaction is quite natural from a qualitative point of view. Such phenomena are well known in classical electrodynamics: an external charge placed in a medium surrounds itself with a polarization cloud, which partially screens it. Therefore, the effective value of the charge observed at large distances is smaller than the true value. When this particle moves its polarization cloud moves with it, leading to an effective change of its inertial properties, i.e., its mass.

In this classical example the particle mass and charge change by a finite amount. The direct application of this physical interpretation of the renormalization equations to quantum field theory is complicated by the singular nature of these equations. This difficulty can be evaded formally by using criteria analogous to those applied in quantum mechanics. The point is that the quantities m_0, α_0, δm, and $\delta\alpha$ entering into the renormalization relations, in contrast to the classical case, cannot be measured separately in experiment, but only in the combinations $Z_m m_0$ and $Z_\alpha \alpha_0$. In contrast to an external charge, which can be removed from the polarized medium and studied in the vacuum, the electron cannot be "freed" from its interaction with the quantum electromagnetic field. Therefore, just as when discussing electron diffraction by a diffraction grating it makes no sense to ask which slit an electron passed through, here it is pointless to make a fuss over the formal nature of the renormalization conditions.

The only thing of physical importance here is the fact that by using the renormalization procedure we arrive at finite, unambiguous expressions which in a number of cases describe the experimental data with an amazing degree of accuracy. For example (see Sec. 16), the value of the electron anomalous magnetic moment calculated in quantum electrodynamics coincides with the experimental value with an experimental error of order 10^{-10}.

12. Renormalization theory

The classification of divergences

The key feature in realizing the renormalization program for a specific quantum field model is the analysis and classification of the types of ultraviolet divergence.

For example, in QED the matrix elements corresponding to diagrams with two and three external lines, shown in Figs. 7, 8, and 9, turn out to be divergent. It can be shown that diagrams with more than three external lines correspond to contributions to the matrix elements which are convergent "as a whole" (the divergences in them arise only from subdiagrams which coincide with one of the diagrams having two or three external lines).

Here the diagrams with two external electron–positron lines, shown in Fig. 7, lead to divergences no worse than linear, diagrams with two external photon lines lead to divergences no worse than quadratic, and diagrams with three external lines lead to logarithmic divergences.

An important feature of QED (and also certain other models of interacting quantum fields, for example, QCD) is the fact that the degree of divergence of a diagram is completely determined by the number and type of external lines and (up to powers of the logarithm $\ln \Lambda$) is independent of the order of the diagram, i.e., the number of internal vertices and loops.

An example of a quantum field model which does not possess this property is the model of a fermion field ψ with a fourfold Fermi self-interaction:

$$\mathscr{L}_{\text{int}} = G\bar{\psi}O\psi\bar{\psi}O\psi.$$

In this model the radiative corrections to the scattering of two fermions on each other correspond to the diagrams shown in Fig. 10. The associated matrix elements diverge as powers. However, in contrast to the earlier case, the degree of divergence depends on the order of the diagrams and increases as the order increases. The leading divergence of an nth-order scattering diagram $G^n\Lambda^{2(n-1)}$ can also be written as $G^{l+1}\Lambda^{2l}$, where l is the number of topologically independent loops of the diagram.

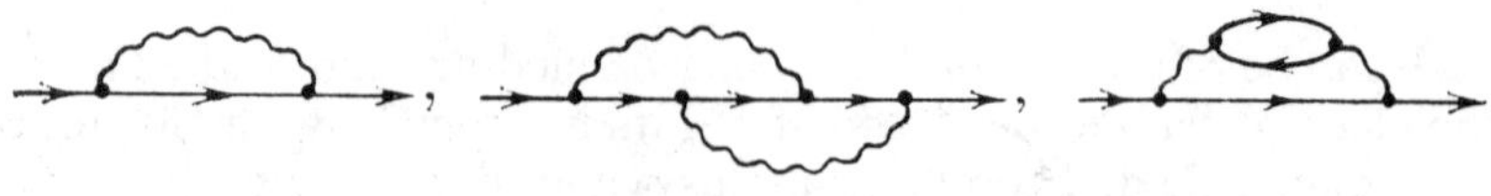

Figure 7. The simplest divergent diagrams with two external electron–positron lines in quantum electrodynamics.

Figure 8. Some of the divergent diagrams with two external photon lines in quantum electrodynamics.

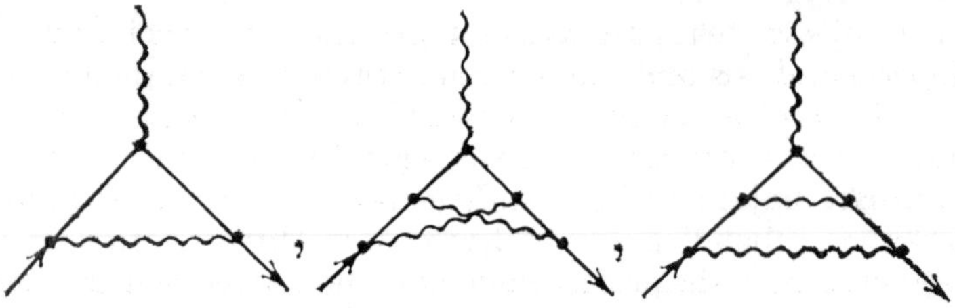

Figure 9. The simplest divergent diagrams with three external lines in quantum electrodynamics.

Quantum field theories such as QED in which the degree of divergence of a diagram is independent of the order of perturbation theory are renormalizable theories. In such models the number of operator structures corresponding to the ultraviolet divergences is finite (and small). For example, in QED there are only two singular operator structures corresponding to the entire infinite set of electron self-energy diagrams (see Fig. 11). Here each of the constants A and B is a series in powers of α with coefficients which diverge for $\Lambda \rightarrow \infty$.

Such operator structures with singular coefficients can be transformed into so-called counterterms, i.e., additions to the Lagrangian of the system of fields. In renormalizable quantum field theories all such additions have the structure of individual components of the full Lagrangian, which, in the final analysis, allows the divergences to be eliminated by redefining the masses and coupling constants.

In models like the 4-fermion model the number of singular structures corresponding to divergent contributions from different orders of perturbation theory is infinite. It is therefore impossible to remove the ultraviolet divergences by the redefinition of a finite number of physical characteristics of the bare particles. Such models are nonrenormalizable.

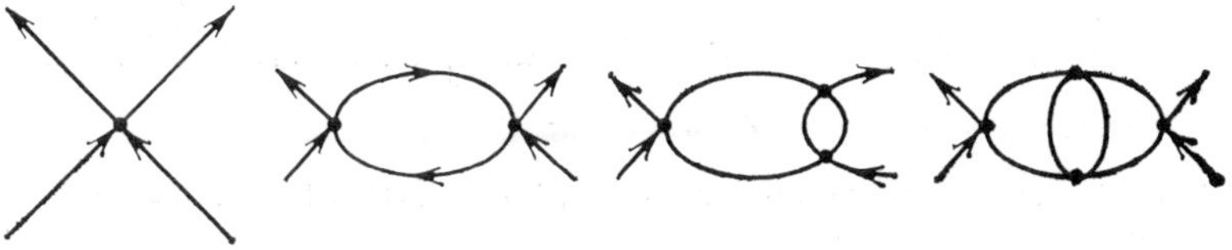

Figure 10. A few of the scattering diagrams in the four-fermion model.

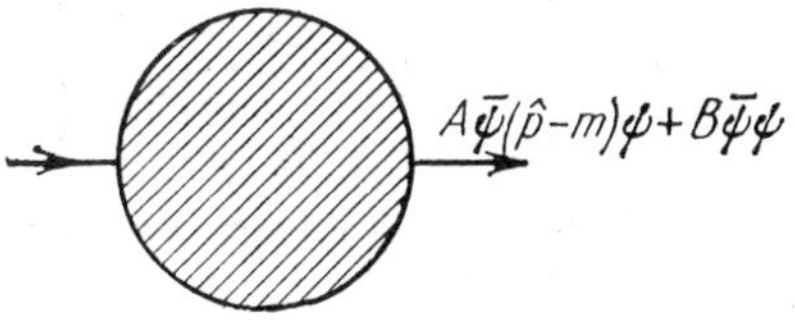

Figure 11. Operator structures with divergent coefficients corresponding to electron self-energy diagrams in quantum electrodynamics.

Renormalizability

Therefore, quantum field models naturally split into two classes—renormalizable and nonrenormalizable models. In renormalizable models it is possible, as a rule, to eliminate all divergences by redefining the parameters (masses, coupling constants,...) of the bare Lagrangian, after which the results of calculations can be expressed uniquely in terms of the renormalized values of the physical parameters.

Without going into the details of the associated formalism, we note that the dimension of the coupling constant can serve as a simple and fairly reliable guide to the renormalizability of a theory. For example, in models where the interaction has the form:

$$g_1\varphi^4(x),\ g_2\bar{\psi}(x)O\psi(x)\varphi(x) \quad \text{or} \quad g_3\bar{\psi}(x)O^\nu\psi(x)B_\nu(x) \tag{12.1}$$

(φ is a scalar, ψ and $\bar{\psi}$ are spinors, and B_ν is a vector), the coupling constants g_i must be dimensionless, since the total dimension of the operators in each of the three expressions is m^4 (or L^{-4}), which coincides with the dimension of the Lagrangian. On the other hand, for the interactions:

$$G\bar{\psi}O\psi\bar{\psi}O\psi,\ H\bar{\psi}O^\nu\psi\frac{\partial\varphi}{\partial x^\nu}, \tag{12.2}$$

in which the total mass dimension of the operators and their derivatives is greater than 4, the coupling constants have negative mass dimension: $[G]=m^{-2}$ and $[H]=m^{-1}$.

In the former case the dimensionless coupling constants are the "natural" expansion parameters of perturbation theory, so the regularized perturbation theory is built up from powers of the coupling constants and products of them and logarithms $g\ln(\Lambda^2/k^2)$ (in the massless case or in the ultraviolet limit). It is therefore quite natural that in such theories the degree of divergence of a graph is, up to a logarithm, independent of the order of perturbation theory. Conversely, for models of the type Eq. (12.2) whose coupling constants have dimensions the natural parameters of the unrenormalized perturbation theory are dimensionless products of the coupling constants and powers of the cutoff momentum: $G\Lambda^2$, $H\Lambda$. In this case the degree of divergence grows with the order of perturbation theory and renormalization is impossible.

This simple picture is not valid in some cases. For example, if in the third of the expressions in Eq. (12.1) the vector field has mass M, it can be used to form a dimensionless expansion parameter, $g\Lambda/M$, and the theory is nonrenormalizable. In the real world this possibility can apparently be neglected, since the known massive vector fields (the W and Z bosons) have a gauge origin. Gauge fields "start out" massless and acquire a mass as a result of spontaneous breakdown of the gauge symmetry (see Sec. 15 below); here the renormalizability is not spoiled. Therefore, a dimensionless coupling constant is a necessary and practically sufficient condition for renormalizability.

What is the physical meaning of the renormalizability condition? To answer this question, we note that radiative corrections can be viewed as effects due to the reaction of the quantum vacuum on the physical particles passing through it. Here we should recall that in quantum field theory the vacuum state is not the physical vacuum. It contains vacuum fluctuations—virtual particles. These vacuum fluctuations—zero-point oscillations—are a well known feature of the lowest state in any quantum theory. In QED the zero-point oscillations are mainly electron–positron pairs and photons which live for a short time. These virtual structures are a sort of "twentieth-century ether" with which physical particles moving through the "vacuum" interact. The ultraviolet divergences of radiative corrections are related to the fact that, as a rule, short-wavelength vacuum fluctuations are very intense. In renormalizable theories their effect can be reduced to a change of the physical parameters of the particles. From this point of view, "renormalizability" means that for the corresponding quantum field interaction mechanisms the effect of short distances (where the high-frequency fluctuations are concentrated) on the physics at large distances can be effectively taken into account by means of a limited number of finite parameters.

It can also be stated that nonrenormalizable interaction models, such as (12.2), correspond to cases where it is not possible to construct a consistent quantum field model with the given interaction mechanism. Since the experience we have gained so far tells us that the property of renormalizability is unrelated to the use of perturbation theory, we can conclude that such models cannot be quantized at all. Therefore, "nonrenormalizability" is a synonym of "nonquantizability."

Renormalization without counterterms

The procedure described above of eliminating ultraviolet divergences by introducing singular counterterms does not reveal the mathematical origin of the infinities. In relation to this, it should be noted that an important stage in the development of renormalization theory was the recognition of the fact that the propagation functions of quantum field perturbation theory are special mathematical objects: they are generalized functions, defined, like the delta function, only as kernels of linear functionals. The operation of multiplying these objects together does not follow from their definition and requires special consideration. As mentioned above in Sec. 5, this view of infinities and their renormalization was developed in the mid-1950s and led to the Bogolyubov–Parasiuk theorem, which provided a rigorous and exhaustive solution of the problem of eliminating divergences.

A consequence of this theorem which is prescriptive in nature and important for applications is the so-called Bogolyubov R-operation. It amounts to the following. To eliminate ultraviolet divergences, instead of introducing some regularization, for example, a cutoff, and manipulating infinite counterterms, it is sufficient to redefine divergent Feynman integrals over virtual momenta by subtracting from them certain simple expressions which are polynomials in the external momenta and in the simplest cases reduce to terms in the Taylor series expansion of the integrand. Uniqueness of the calculations is ensured by imposing special conditions on the finite, i.e., now divergence-free, results of the calculations. As a rule, these conditions can be chosen such that the Lagrangian parameters coincide with the physical ones.

Surface divergences

It is important to stress the fact that the renormalization procedure described above eliminates ultraviolet divergences only from the scattering matrix, which, as mentioned above in Sec. 10, is obtained by a special limit as $t \to \infty$ from the solution of the Schrödinger equation in the interaction representation. As far as the solution itself, i.e., the state vector in the interaction representation $\Psi(t)$ at some infinite time t, is concerned, here the situation remains unsatisfactory even after renormalization.

The point is that the changeover to $\Psi(t)$ requires abandoning the picture of the smooth (adiabatic) switching on and off of the interaction for $t \to \pm \infty$, which is sufficient for introducing the S-matrix. Instead, we must deal with the situation where the interaction is switched off sharply at time t (in general, on some spacelike surface). Detailed analysis shows that in this case counterterms containing derivatives of the fields lead to the appearance in $\Psi(t)$ of special, so-called surface divergences, the existence of which was first pointed out by Stueckelberg.

The presence of surface divergences makes it impossible to obtain the state vector in the interaction representation. This fact is associated with the Haag theorem mentioned in Sec. 7.

13. Renormalization group

Effective coupling constants

One of the observable manifestations of vacuum fluctuations in QED is the partial screening of the electron charge. The point is that in order to measure the charge, the electron must be placed in an electromagnetic field, and then the reaction of the field must be studied. In other words, the charge must be "probed" using the quanta of this field—photons. Here it turns out that on its way to the charge a quantum probe can undergo virtual dissociation into an electron–positron pair, which forms an effective dipole leading to the screening effect. This process gives a contribution which is proportional to α and depends on the photon 4-momentum or, in the coordinate representation, the distance to the measured charge. Therefore, the inclusion of vacuum effects transforms a numerical parameter—the electron charge—into a function referred to as the effective charge.

The behavior of the effective charge in QED qualitatively corresponds to the classical screening of an external charge placed in a polarizable medium: its value decreases with increasing distance. In the 4-momentum representation commonly used in quantum field theory, this corresponds to a function $\overline{\alpha}(Q^2)$ which increases monotonically with Q^2. The dashed lines in Fig. 12 show the dependences $\overline{\alpha}(Q^2)$ which are theoretically possible *without* regard to the experimental data. For QED, experiment indicates that $\overline{\alpha}(0) = 1/137$, thereby selecting one of the possible curves for the effective charge.

We note that, in general, in any renormalizable quantum field model the quantum corrections cause the constant numerical parameter g in the bare Lagrangian to become a function $\bar{g}(Q^2)$, called the effective coupling constant. One of the possible curves in each case is selected using the condition $\bar{g}(\mu^2) = g_\mu$. The results of renormalized perturbation theory are therefore naturally parametrized by the pair μ and g_μ. The appearance of the new parameter μ, the scale parameter, in the theory is referred to as dimensional transmutation.

Renormalization invariance

From the discussion of the effective coupling constant in QED it follows that the physical interpretation of the scale parameter μ is that it characterizes the measurement conditions (for example, the photon momentum) which lead to the experimentally measured value g_μ of coupling constant. If the measurement conditions are changed, we find a different value of the effective coupling constant, with the new pair of values of μ and g_μ lying on the same curve as the old pair. It is when this condition is satisfied that the two parametrizations are physically equivalent.

Let us now consider in more detail the operation of going from one parametrization, for example, (μ_1, g_1) to another, (μ_2, g_2) (see Fig. 13). For simplicity we assume that we are dealing with a massless quantum field model, for

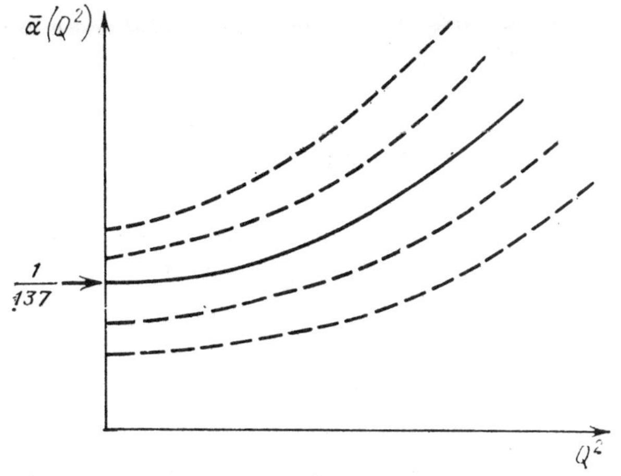

Figure 12. Qualitative behavior of the effective charge in QED as a function of the squared 4-momentum. The solid line is fixed by agreement with the Millikan experiment.

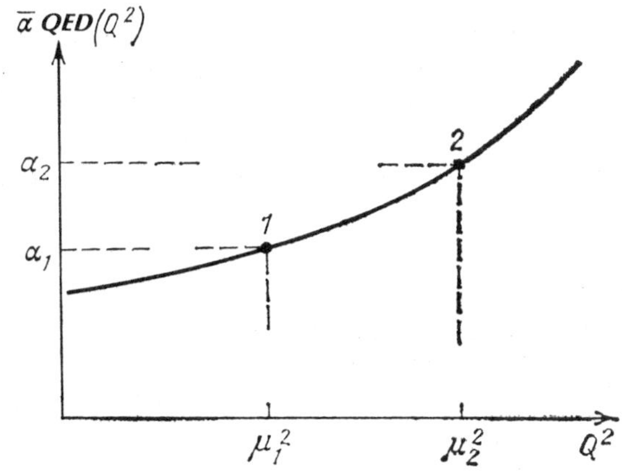

Figure 13. The possibility of different parametrizations of actual quantum electrodynamics corresponding to the Millikan experiment. Passage along the curve from point 1 to point 2 corresponds to a renormalization group transformation.

example, massless QED (which physically corresponds to the ultraviolet limit of ordinary QED).

Here after the normalization condition is imposed the "physical" curve $\bar{g}(Q^2)$ can be represented as a function of two dimensionless arguments Q^2/μ^2 and g_μ, i.e., it can be written as $\bar{g}(Q^2/\mu^2, g_\mu)$. However, the pair of quantities μ and g_μ used in this parametrization can correspond to any point on the selected curve. The condition of renormalization invariance means that any two different parametrizations, for example, (μ_1, g_1) and (μ_2, g_2) must lead to the same result, in particular, for the function $\bar{g}$ itself, i.e., we must have

$$\bar{g}\,(Q^2/\mu_1^2, g_1) = \bar{g}\,(Q^2/\mu_2^2, g_2), \tag{13.1}$$

where obviously $g_2 = \bar{g}\,(\mu_2^2/\mu_1^2, g_1)$ and $g_1 = \bar{g}\,(\mu_1^2/\mu_2^2, g_2)$. Combining these expressions we arrive at the functional equation:

$$\bar{g}\,(x, g) = \bar{g}\,[x/t, \bar{g}\,(t, g)]. \tag{13.2}$$

This equation, together with the corresponding expressions relating the explicit form of the Green functions and the matrix elements for different but physically equivalent parametrizations, provides the basis for the mathematics of the renormalization group method.

The renormalization group

The renormalization group can be defined as the group of transformations from one of the possible parametrizations to another, physically equivalent one. These transformations form a continuous, one-parameter group, which can also be viewed as the group of scale transformations of the squared momenta by the factor t with the simultaneous functional transformation of the coupling constant:

$$x \to x' = x/t, \quad g \to g' = \bar{g}\,(t, g). \tag{13.3}$$

The functional equation (13.2) ensures the group composition rule for Eq. (13.3).

The fundamental application of the Stueckelberg–Petermann renormalization group in quantum field theory is to the partial summation of the logarithmic contributions to the ultraviolet asymptote. This is accomplished by a special procedure known as the renormalization group method, based on the use of the group differential equations. Setting this application aside for now (it will be discussed later in this section), we note that the renormalization group was later used successfully in the theory of critical phenomena, the theory of turbulence, the theory of radiation transfer, and in several other areas of physics far removed from the microworld and quantum phenomena. The extension of the renormalization group to areas outside quantum field theory is largely due to K. Wilson.

The reason for this universality lies in the simplicity and generality of transformations like Eq. (13.3). As has recently been shown by one of the present authors, these transformations reflect the transitive invariance of simple physical systems with respect to the specific manner in which they are parametrized using initial or boundary values of suitable physical quantities. At the time they represent a simple generalization of the power-law similarity transformations (scaling transformations) known more than a hundred years ago in fluid mechanics and gas dynamics and today widely used in a wide variety of problems in mechanics and physics.

The renormalization group method

In quantum field theory this is a systematic method of "improving" the results of ordinary perturbation theory. The point is that the exact solutions of the quantum field equations must satisfy the condition of renormalization invariance. In practice we deal with pieces of Taylor series expansions in the coupling constant. The properties of these approximate solutions can differ significantly from those of the exact solutions because renormalization invariance is violated in perturbation theory. These differences become important when the solution has a singularity, as occurs, for example, in the ultraviolet limit. The renormalization group method allows one to obtain from the approximate result a renormalization-invariant expression which coincides with the original expression in certain lowest orders in g. The group differential equations serve as the technical tool for realizing this program. The simplest (and, perhaps, most important) of these equations can be obtained by differentiating the functional equation (13.2). It has the form:

$$x\,\frac{\partial \bar{g}\,(x,g)}{\partial x} = \beta\,[\bar{g}\,(x,g)]. \tag{13.4}$$

It is easy to show that for any given function $\beta(g)$ the solution of this equation exactly satisfies Eq. (13.2). The beta function on the right-hand side can be expressed in terms of the solution $\bar{g}$ as

$$\beta(g) = \left.\frac{\partial \bar{g}\,(x,g)}{\partial x}\right|_{x=1}. \tag{13.5}$$

The renormalization group method amounts to obtaining the beta function from a given approximate solution using Eq. (13.5). Then the solution of Eq. (13.4) gives an improved approximate solution which, on the one hand, corresponds to the original solution in the lowest orders and, on the other, is renormalization-invariant. The asymptotic properties are changed significantly as a result of this procedure.

In QED the one-loop approximation for $\bar{\alpha}$ gives

$$\bar{\alpha}^{(1)}(x,\alpha) = \alpha + \frac{\alpha^2}{3\pi}\ln x.$$

Using this expression to find $\beta(\alpha) = \alpha^2/3\pi$ and solving the differential equation (13.4), we obtain the improved renormalization group solution:

$$\bar{\alpha}(x,\alpha) = \frac{\alpha}{1 - \dfrac{\alpha}{3\pi}\ln x}. \tag{13.6}$$

The analogous expression for the effective expansion parameter in QCD:

$$\bar{\alpha}_s(x,\alpha) = \frac{\alpha_s}{1 + \alpha_s\,\beta_1\,\ln x}, \quad \beta_1 = \frac{33 - 2n}{12\pi}, \tag{13.7}$$

(where n is the number of quark flavors) qualitatively differs from the behavior of $\bar{\alpha}$ in QED because the coefficient β_1 is positive.

Asymptotic freedom and the "zero charge"

Equation (13.7) describes the phenomenon of asymptotic freedom, namely, that the interaction strength described by the effective coupling constant decreases with decreasing distance (or with increasing Q^2).

Discovered theoretically in the gauge model in which the strong interaction between quarks is transmitted by quanta of the non-Abelian Yang–Mills field, the phenomenon of asymptotic freedom allowed a quantum field interpretation of the parton model of hadron structure. Moreover, it proved to be the decisive argument in the formulation of the quantum field theory of the strong interaction, now known as quantum chromodynamics. Equation (13.7) can therefore be viewed as the most important physical application of the renormalization group method.

We see that the coefficient β_1 in Eq. (13.7) has two terms. The one proportional to n comes from internal fermion loops and, since it has the same sign as the one-loop contribution in Eq. (13.6), it favors the growth of $\overline{\alpha}$ with increasing x. The second, positive term $11/4\pi$ is related to purely gluonic effects, i.e., to diagrams which do not contain virtual fermions. Asymptotic freedom therefore arises entirely from nonlinear effects of the Yang–Mills quantum field itself.

The quantum electrodynamical equation (13.6) received a great deal of attention from theoreticians in the mid-1950s. The reason was that, if viewed as being exact, this equation has a feature which can be interpreted as an indication that local QED is inherently inconsistent. This equation contains a pole at $x = x_* = \exp(3\pi/\alpha)$ which does not admit any reasonable physical interpretation owing to the negative sign of the residue. It is therefore said to be a "ghost" pole. In the naive formulation of the analogous result in the language of cutoff momenta, one finds formally that the renormalized electron charge vanishes when the bare charge is finite and the cutoff momentum Λ is taken to infinity. This situation is widely referred to as the "zero-charge problem."

At the time, this difficulty gave rise to serious doubts about the self-consistency of quantum electrodynamics and also local quantum field theory in general. As noted above (at the end of Sec. 5 and the end of Sec. 6), the zero-charge phenomenon together with the unsuccessful attempts to apply the Feynman diagram technique based on the Yukawa interaction mechanism to problems in hadronic physics led to a great deal of skepticism among physicists regarding quantum field theory as the basis for the theoretical description of particle interactions.

However, more careful analysis shows that the zero-charge problem was greatly overemphasized. Already in 1955 Bogolyubov and one of the present authors showed that there was no conclusive evidence of the presence of a contradiction in quantum electrodynamics. Incidentally, we note that a simple analysis of the ghost pole problem can be carried out using the renormalization group method. The crucial feature is that Eq. (13.6) is not exact. It corresponds only to the leading logarithm approximation, i.e., it represents the sum of iterations of one-loop contributions of the form $\alpha(\alpha \ln x)^n$. It can be improved by including the two-loop contributions in the determination of the beta function using perturbation theory. After solving the differential equation, one obtains an expression more complex than Eq. (13.6) which effectively includes also the nonleading logarithmic contributions of the form $\alpha^2(\alpha \ln x)^n$. This equation from the two-loop renormalization group approximation differs significantly from Eq. (13.6) in the region $x \sim x_*$. The inclusion of still higher orders (three-loop contributions and so on) changes the picture in this region even more. It is therefore impossible to make any definite statements about the existence of a pole.

The current point of view is that a possible (in principle) problem in local QED at distances of order 10^{-280} cm cannot be of any physical significance, because it is widely believed that the interactions are unified at distances of order 10^{-29}–10^{-33} cm (see Sec. 22 below).

Here we should also note that in 1955 in connection with the zero-charge problem Landau considered the possible unification of gravity and electromagnetism at the Planck scale ($\sim 10^{-33}$ cm).

This unification was later studied by Markov, who showed that the inclusion of the gravitational interaction leads to an upper limit on the masses of intermediate states in QED. This maximum mass of 10^{-5} g, called the "friedmon" or the "maximon," provides an effective cutoff at the Planck scale.

Chapter 4
Local symmetries and their breakdown

14. Gauge fields

Local phase transformations

In Sec. 6 we discussed how internal symmetries are used for hadron systematics. However, symmetries also play other roles in particle physics. The concept of internal symmetry is of fundamental importance for understanding the nature of interactions between particles.

As is well known, in quantum mechanics the states of a system are described by a wave function (state vector) Ψ. The observable expectation values of physical quantities and the amplitudes for transitions of the system from one state to another are represented as bilinear combinations $\overset{*}{\Psi}_2 O \Psi_1$, where O is some operator acting on the wave functions. Since the wave functions are complex-valued functions characterized by a modulus and a phase, it follows that only phase differences, rather than the phases themselves, have a physical meaning.

In quantum field theory the state vectors are obtained by allowing operator functions of the field to act on the vacuum vector. Therefore, the phases of field functions also do not have any real meaning; only phase differences are important. A change of the phase of a field function by an arbitrary additive quantity:

$$\varphi(x) \equiv |\varphi(x)| e^{i\alpha(x)} \to \varphi'(x) \equiv |\varphi(x)| e^{i(\alpha(x)+\lambda)} \tag{14.1}$$

therefore cannot lead to a change in the physics.

This implies that the Lagrangian of a system of fields must be invariant under such phase transformations, i.e., it must have the same form when written in terms of the fields $\varphi(x)$ and $\varphi'(x)$.

The invariance of the Lagrangian under transformations (14.1) with $\lambda =$ constant implies the existence of a conserved characteristic of the field identified as the charge. This property is a consequence of Noether's first theorem, which associates continuous symmetry transformations with certain quantities conserved over time. A consequence of this theorem is, for example, the law of energy–momentum conservation, which is related to translational invariance in spacetime.

Phase transformations of field functions can be viewed as transformations in some internal (or, alternatively, charge) space. In this case, the charge space is the complex plane, and it is easy to see that the transformations form the group U(1). Later we will encounter more complicated internal spaces.

Phase transitions (14.1) with $\lambda = \text{const}$ are referred to as global. They can be used to change the phase of the field at some arbitrary point. However, if the phase of the field at some point in spacetime is fixed in this manner, the value of the phase will thereby be fixed at all the other points, since global phase transitions are identical for all x. But actual experiments are always carried out in limited regions of spacetime. It is therefore natural to be able to choose the phase independently at different points. For this we must require that the theory be invariant also when the transformation parameter depends on x:

$$\varphi(x) \to \varphi'(x) = \varphi(x) e^{i\lambda(x)}, \tag{14.2}$$

i.e., that the relative phase $\lambda(x) - \lambda(x')$ at two different spacetime points x and x' be completely arbitrary. Such phase transformations are referred to as local.

In the case of free fields, however, there is invariance only under global transformations. Under local transformations the derivatives of the fields $\varphi(x)$ and $\varphi'(x)$ differ by an additive quantity proportional to the gradient of the function $\lambda(x)$. The Lagrangian of the free field φ and the equations of motion obtained from it, which contain the term $\partial_\mu \varphi$ $(\partial/\partial x^\mu \equiv \partial_\mu)$, therefore are not invariant under (14.2).

The electromagnetic field

Invariance under local transformations can be obtained by introducing into the theory an auxiliary vector field $A_\mu(x)$, which transforms simultaneously with a phase transformation of the field $\varphi(x)$ according to the rule:

$$A_\mu(x) \to A'_\mu(x) = A_\mu(x) + \frac{1}{e}\frac{\partial \lambda(x)}{\partial x^\mu}. \tag{14.3}$$

Here all derivatives $\partial\varphi/\partial x^\mu$, $\partial\overset{*}{\varphi}/\partial x^\mu$ must be replaced by the operators

$$D_\mu(A)\varphi \equiv \left(\frac{\partial}{\partial x^\mu} - ieA_\mu(x)\right)\varphi(x);$$

$$\overset{*}{D}_\mu(A)\overset{*}{\varphi} \equiv \left(\frac{\partial}{\partial x^\mu} + ieA_\mu(x)\right)\overset{*}{\varphi}(x), \tag{14.4}$$

referred to as "extended" derivatives. They are also called "covariant," since under the simultaneous transformation of the fields φ, $\overset{*}{\varphi}$, and A_μ they transform like the fields φ and $\overset{*}{\varphi}$, i.e., they change phase:

$$D_\mu(A')\varphi' = e^{i\lambda(x)} D'_\mu(A)\varphi(x),$$
$$\overset{*}{D}_\mu(A')\overset{*}{\varphi}(x) = e^{-i\lambda(x)} \overset{*}{D}_\mu(A)\overset{*}{\varphi}(x).$$

From this we easily see that a Lagrangian containing the fields φ and $\overset{*}{\varphi}$ and their covariant derivatives is invariant under the transformations (14.2) and (14.3).

Figure 14. Vertices for the interaction between scalar and electromagnetic fields.

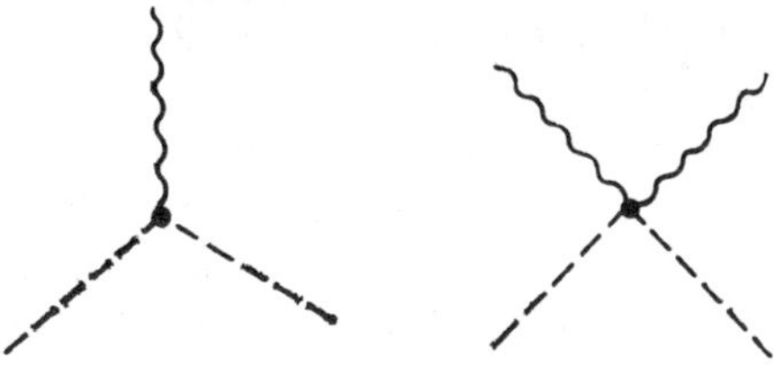

The replacement of ordinary derivatives by covariant derivatives in the free Lagrangian gives rise to an interaction between the field $\varphi(x)$ and the vector field $A_\mu(x)$. To complete the picture we must add the Lagrangian of the free field $A_\mu(x)$, which in turn must be invariant under the transformations (14.3). It has the form

$$\mathscr{L}_0(A)' = -\tfrac{1}{4} H_{\mu\nu} H^{\mu\nu}, \text{ where } H_{\mu\nu} = \frac{\partial A_\nu}{\partial x^\mu} - \frac{\partial A_\mu}{\partial x^\nu}. \tag{14.5}$$

If we identify the constant e appearing in Eqs. (14.3) and (14.4) as the electric charge, the vector field $A_\mu(x)$ is the electromagnetic field. Here the antisymmetric tensor $H_{\mu\nu}$ has a simple expression in terms of the components of the three-dimensional electric and magnetic field strength vectors:

$$H_{\mu\nu} = \begin{pmatrix} 0 & -E_1 & -E_2 & -E_3 \\ E_1 & 0 & H_3 & -H_2 \\ E_2 & -H_3 & 0 & H_1 \\ E_3 & H_2 & -H_1 & 0 \end{pmatrix}. \tag{14.6}$$

We note that the introduction of the interaction by means of covariant derivatives corresponds exactly to the method used in classical electrodynamics, as discussed in Sec. 4. As is well known, the theory obtained in this manner possesses the property of gauge invariance.

Since the Lagrangian of a free spinor field is linear in the derivatives, the resulting electromagnetic interaction is linear in the field A_μ. The interaction Lagrangian has the form:

$$e\bar{\psi}(x)\gamma^\mu\psi(x)A_\mu(x).$$

The Lagrangian of a free scalar field is quadratic in the derivatives. The interaction is therefore described by terms linear and quadratic in A_μ: $ie[\overset{*}{\varphi}\,\partial\varphi/\partial x^\mu - \partial\overset{*}{\varphi}/\partial x^\mu\,\varphi\,]A_\mu$, and $e^2\overset{*}{\varphi}\varphi A_\mu A^\mu$. The corresponding vertices are shown in Fig. 14.

The invariance of the theory under the transformations (14.2) and (14.3) is known as (local) gauge invariance, and the auxiliary vector field which was introduced in order to cancel the local change of phase (gauge) of the field φ is called a gauge field. Furthermore, in order to emphasize the special role of the gauge fields, the other fields are often called matter fields, in spite of the fact that all fields are, obviously, "matter" fields in the same sense.

The group of phase transformations of the complex matter field $\varphi(x)$ is Abelian. This feature is also reflected in the gauge field terminology. For example, the electromagnetic field $A_\mu(x)$ is an Abelian gauge field. In the case of multicomponent matter fields there can be invariance under local transformations forming a non-Abelian group. Here it is necessary to introduce non-Abelian gauge fields, also called Yang–Mills fields.

Yang–Mills fields

Let a multicomponent field function $\varphi_a(x)$ $(a = 1,..., A)$ transform according to a representation of some non-Abelian Lie group G. It is sufficient to consider only infinitesimal transformations, which completely define a Lie group. An infinitesimal local gauge transformation of the field φ_a looks like

$$\varphi_a \to \varphi'_a = \varphi_a + [\lambda(x)]_{ab}\varphi_b. \tag{14.7}$$

The matrix $\lambda(x)$ entering into this expression is expressed as a linear combination of the matrices T_k, the generators of the given representation, with infinitesimal coefficients $\alpha_k(x)$ depending on the space-time point:

$$\lambda(x) = ig\, \Sigma_k \alpha_k(x)\cdot T_k.$$

To make the Lagrangian of the free field $\varphi_a(x)$ invariant under the transformations (14.7), by analogy with electrodynamics we replace the ordinary derivatives $\partial\varphi_a/\partial x^\mu$ by covariant derivatives $D_\mu(B)\varphi_a$ containing new vector fields $B_\mu^k(x)$ whose transformation leads to cancellation of the additive term arising from differentiation of $\lambda(x)$.

In contrast to the electromagnetic field, these fields carry the group index k. It is easy to see that the maximum number of independent fields $B_\mu^k(x)$ is equal to the dimension of the group G $[N^2 - 1$ for $\mathrm{SU}(N)]$.

The fields $B_\mu^k(x)$ transform according to a more complicated law than that for the electromagnetic field. In addition to the "gauge extension" $\partial\lambda(x)/\partial x^\mu$ they contain a term corresponding to group rotation analogous to Eq. (14.7).

The analog of the electromagnetic field strength tensor $H_{\mu\nu}$ is the tensor:

$$F_{\mu\nu} = \frac{\partial B_\nu}{\partial x^\mu} - \frac{\partial B_\mu}{\partial x^\nu} + ig(B_\mu B_\nu - B_\nu B_\mu); \tag{14.8}$$

here B_μ is the linear combination $B_\mu(x) = \Sigma_\mu B_\mu^k(x) t_k$, where the matrices t_k are the generators of the adjoint representation of the group G. In contrast to $H_{\mu\nu}$, the tensor $F_{\mu\nu}$ is not itself invariant. The transformation of $F_{\mu\nu}$ takes the form of rotations in the space of the adjoint representation. However, the Lagrangian of the field B_μ:

$$\mathscr{L}_0(B) = \mathrm{Tr}\, F_{\mu\nu} F^{\mu\nu}$$

(Tr stands for the trace of the matrix) is invariant, as can easily be shown.

Just like the electromagnetic field $A_\mu(x)$, the fields $B_\mu(x)$ are necessarily massless, since the Lagrangians cannot contain noninvariant terms of the type $m^2 B_\mu B^\mu$ describing the vector field mass.

The requirement that the gauge fields be massless greatly restricted the construction of models of real interactions involving gauge fields, because,

Figure 15. Self-interaction vertices of a non-Abelian gauge field.

aside from the electromagnetic field, no other massless vector fields are observed in experiment. One possible way to avoid this difficulty is to assume that gauge fields are unobservable in principle. This is essentially what is done in quantum chromodynamics, the current version of the strong interaction theory. Another possibility is to introduce a vector field mass through some "soft" breaking of the gauge symmetry while preserving many of the features of the invariant theory, in particular, its renormalizability. The following section is devoted to the description of a mechanism for such symmetry breaking.

An important feature of the Lagrangian of a non-Abelian gauge field is the presence of terms proportional to

$$\mathrm{Tr}\left[B_\mu B_\nu\left(\frac{\partial B_\mu}{\partial x^\nu}-\frac{\partial B_\nu}{\partial x^\mu}\right)\right] \text{ and } \mathrm{Tr}[B_\mu B_\nu B_\mu B_\nu],$$

which correspond to the interaction of the field $B_\mu(x)$ with itself. They are associated with the diagrams with three- and fourfold vertices shown in Fig. 15.

The interaction of non-Abelian gauge fields with matter fields has the same form as the electromagnetic field interaction, differing only in the multiplication of the vertices by certain constant matrices—the generators of the representations corresponding to the matter fields.

We note that the arbitrariness in the choice of local gauge parameter $\lambda(x)$ means that, as in electrodynamics, of the four components of the vector field $B_\mu(x)$, only the two "spatially transverse" components correspond to physical degrees of freedom. A feature of the non-Abelian gauge field which greatly complicates its study is that the self-interaction causes the physical degrees of freedom to mix with the unphysical, longitudinal, ones. The latter therefore are also important in actual calculations.

This difficulty made it necessary to develop special methods for quantizing gauge fields, which can be viewed as dynamical systems with constraints. In quantizing such systems it is necessary to separate the true dynamical variables from the unphysical ones, while eliminating as many pairs of canonically conjugate variables as there are constraints. The formalism most convenient for gauge field quantization is that of the path integral.

The path integral

The concept of the path, or, equivalently, the functional integral was introduced in quantum theory by Feynman in 1948. Feynman's starting point was the idea, which originated earlier with Dirac, that the time evolution of a quantum system from the instant 0 to time T can be represented as a sequence of evolutions in a large number of small time intervals $(0,t_1)$, $(t_1,t_2),\ldots,(t_N,T)$. If we then consider the corresponding transformation function in the representation in which the coordinate q is the observable, i.e., the quantity $\langle q(T)|q(0)\rangle$, it proves possible to represent it as a multiple integral of the product $\langle q(T)|q_N\rangle\cdots\langle q_2|q_1\rangle\langle q_1|q(0)\rangle$ over the possible values of the coordinate $q_1,q_2,\ldots,q_N$ at the intermediate times $t_1,t_2,\ldots,t_N$. By shrinking the time intervals while taking the number N of integrations to infinity, we arrive at a new representation of quantum mechanical quantities in terms of path integrals. From the formal point of view, the path integral is the limit of a multiple integral as the number of integrations tends to infinity. In the quantum mechanical case discussed above we obtain an expression which can be treated as an integral over all possible values of the particle coordinates in the time interval from 0 to T, i.e., over all possible values of its trajectory during this time interval. The path integral is thereby represented as an integral over all possible values of the function $q(t)$. Such integrals are therefore also known as functional integrals.

In field theory the canonical coordinates are the values of the field functions at all possible points in 3-dimensional space. The path integral in quantum field theory is therefore a functional integral over the field functions.

Feynman first used the path integral to construct quantum field perturbation theory, but it later became clear that the representation of quantum field quantities in terms of functional integrals is also very useful for calculations outside the framework of perturbation theory (see Sec. 21).

It was found that this new object possesses the algebraic properties of an ordinary integral and permits integration by parts and substitution of the integration variables. The path integral formalism is therefore very useful for studying the effects of transformations, including gauge transformations, of the field functions. This particular feature accounts for its special role in the quantization of gauge fields.

Both in quantum mechanics and in quantum field theory, path integrals can be interpreted as expressions for averaging over the quantum fluctuations around solutions of the classical equations of motion. This viewpoint allows the path integral representation to be viewed as an alternative to other quantization methods.

The Yang–Mills quantum field

In the quantization of gauge fields, which are systems with constraints, the functional integral representation permits these constraints to be taken into account relatively simply, by introducing the corresponding restrictions in the integration in the function space of the components of the gauge vector field $B_\mu^a(x)$. The resulting path integral can, after some transformations, be written in the standard form for generating the diagrams of quantum perturbation theory. This corresponds to the following effective interaction Lagrangian for the quantum field B_μ^a:

$$\mathscr{L}_{\text{eff}} = \frac{g}{2}(\partial_\nu B_\mu^a - \partial_\mu B_\nu^a)B_\mu^b B_\nu^c t_{abc}$$

$$- \frac{g^2}{4} B_\mu^a B_\nu^b B_\mu^d B_\nu^e t_{abc} t_{cde} + g\overline{C}^a(x) B_\nu^b(x) \partial^\nu C^c(x) t_{abc}$$

(t_{abc} are the structure constants—the matrix elements of the generators of the adjoint representation; see Sec. 3). In addition to terms of higher order in B and ∂B, this also contains a term corresponding to the interaction of the field B_μ with the scalar Fermi field of the "Faddeev–Popov ghosts" $C(x)$, which is absent in the Lagrangian of the classical Yang–Mills field. Accordingly, the Feynman diagrams and rules involve, in addition to the propagator of the field B_μ and its self-interaction vertices, the propagator of the "ghost" field C and the vertices of its interaction with the vector field B_μ.

Here the diagrams of perturbation theory contain ghost particles only in intermediate states in the form of closed loops. The role of the contributions from such loops is to cancel the corresponding contributions from the unphysical components of the field B_μ. They therefore restore S-matrix unitarity in the space of physical states.

The interaction between the gauge field B_ν and the matter fields is described by the corresponding vertices obtained by the appropriate extension of the derivatives of these fields.

15. Spontaneous symmetry breaking

Spontaneous symmetry breaking is a phenomenon which is well known in diverse areas of physics. It amounts to the following: a physical system described by equations of motion which possess a certain symmetry is in a state corresponding to a particular solution of the equations of motion which does not possess this symmetry.

Spontaneous symmetry breaking occurs in those cases where the system does not possess the absolute minimum energy in the symmetric state with the lowest energy, so that this state is unstable. The actual cause of the symmetry breaking is an arbitrarily small asymmetric perturbation.

An example from mechanics

This phenomenon is clearly illustrated by the following system. A symmetrically shaped container (a body of revolution) with a convex bottom is placed in the gravitational field such that the symmetry axis of the container is parallel to the gravitational force. A small ball is also located on this axis (see

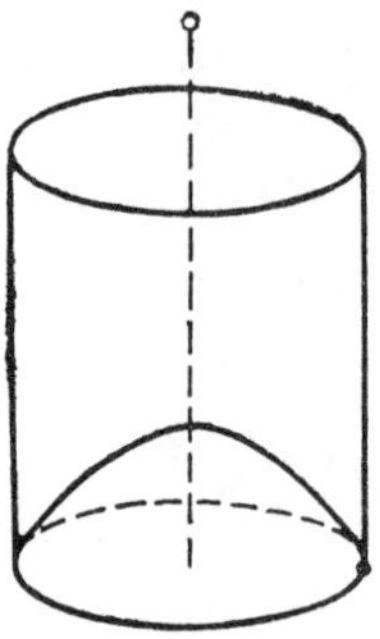

Figure 16. A mechanical system demonstrating spontaneous symmetry breaking.

Fig. 16). If the ball is dropped along the axis, it first hits the bottom of the container exactly in the middle, but since the center of the convex bottom corresponds to a position of unstable equilibrium, the ball rolls down to the edge of the bottom at some point. At the initial instant of time the system was symmetric under rotations about the container axis, but this symmetry is not present in the final state.

Examples from quantum statistics

Another familiar example is a ferromagnet. The state which is symmetric under rotations of three-dimensional space, in which the magnetic moments of the various atoms are oriented randomly, is not the state of minimum energy. In a ferromagnet the magnetic moments of neighboring atoms are correlated, so that the energy of the system is smaller in a state with parallel spins. Therefore, in the state of minimum energy the moments of all the atoms are aligned parallel to each other and the material possesses a macroscopic magnetic moment. There are obviously infinitely many such states with minimum energy (differing from each other by rotations of the atomic magnetic moments by the same angle). However, in reality the system is located in one such state.

Another example, which, perhaps, is more subtle but is closer to the situation in quantum field theory models is that of superconductivity. This phenomenon occurs when, owing to certain interactions with lattice vibrations, electrons with oppositely oriented momenta and spins form pairs. These pairs, being bosons (the total spin of a pair is equal to zero), accumulate in the lowest energy state. The symmetry which is broken in this case is the gauge symmetry corresponding to the conservation of the electric charge of a pair. This is related to the fact that the number of pairs, i.e., the number of ordered electrons, is not a given quantity. To characterize the ordering a special parameter is introduced which is analogous to the direction of the magnetic moment of a ferromagnet. The specification of this parameter fixes the arbitrariness.

Bogolyubov developed a special method for analyzing quantum statistical systems with spontaneous symmetry breaking: the method of quasiaver-

Figure 17. Potential function of a free scalar field.

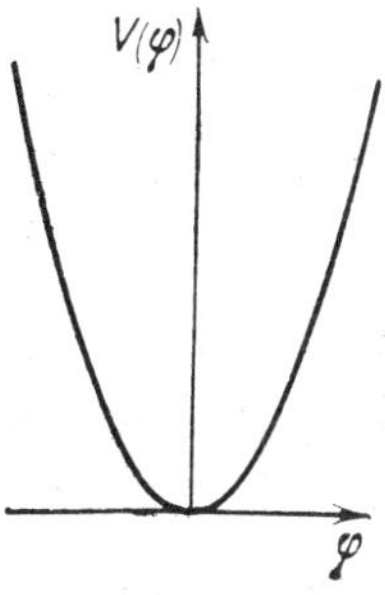

ages. Here the degeneracy in the system is lifted by a certain infinitesimal perturbation, then this perturbation is allowed to tend to zero.

The field theory case

Let us now see how spontaneous symmetry breaking occurs in field theory. Here also there can exist physical states which do not possess the symmetry of the Lagrangian. In particular, the ground state, which has the minimum energy and is naturally identified as the vacuum, may not be symmetric.

The simplest example of a system of this type is a self-interacting scalar field, whose Lagrangian contains a mass term of the "wrong" sign. In general, the Lagrangian of a scalar field $\varphi(x)$ has the form $\mathcal{L} = T(\partial\varphi/\partial x^{\mu}) - V(\varphi)$, where T is the kinetic part, depending only on the components of the gradient $\partial\varphi/\partial x^{\mu}$, and V is the potential function.

For an ordinary free scalar field $V(\varphi)$ is the parabola $m^2\varphi^2/2$ shown in Fig. 17. The minimum of the potential function corresponds to the classical solution $\varphi = 0$, and the quantization of the free scalar field amounts to the quantization of the small oscillations about the equilibrium position at $\varphi = 0$. The Lagrangian for this case is symmetric under reflection $\varphi \to -\varphi$, and the ground state $\varphi = 0$ is symmetric and stable. The introduction into $V(\varphi)$ of positive terms of the form $+\varphi^4$ describing the self-interaction of the field does not change these properties.

In order to spontaneously break the reflection symmetry we assume that the φ dependence of V takes the form of an even function with two symmetric minima, as shown in Fig. 18. This can be written as $V(\varphi) = -(\mu^2/2)\varphi^2 + (h^2/4)\varphi^4$. Owing to the negative sign of the first term, the point $\varphi = 0$ is not a minimum, and there are two positions of stable equilibrium $\varphi = \pm\varphi_0$, where $\varphi_0 = \mu/h$. The two lowest states have the same energy $V(\pm\varphi_0) \equiv V_0 = -(\mu^4/4)h^2$, so the corresponding quantum system has a doubly degenerate vacuum state.

To study small deviations from equilibrium we choose one of the lowest states, for example, $+\varphi_0$, and by a constant displacement transform to a new field function: $\varphi(x) \to \varphi'(x) = \varphi(x) - \varphi_0$. In terms of the new field function V has the form:

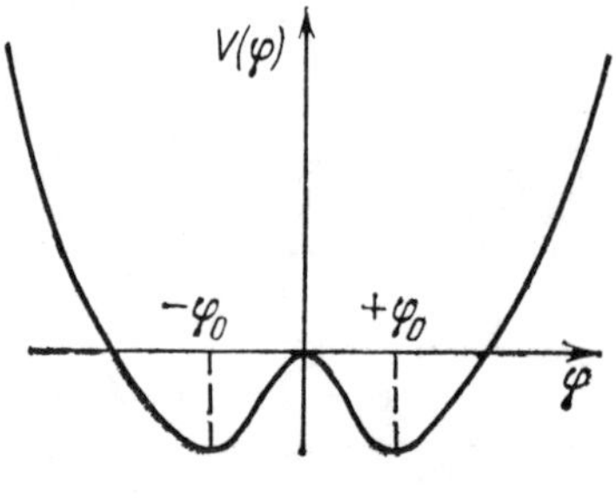

Figure 18. Potential function of a scalar field model with spontaneous symmetry breaking.

$$V = V_0 + \mu^2\varphi'^2 + \mu h\varphi'^3 + \frac{h^2}{4}\varphi'^4.$$

Ignoring the constant V_0, which corresponds only to a shift in the energy scale, we easily see that the positive second term on the right-hand side corresponds to the mass of the small oscillation quanta $m = \sqrt{2}\mu$, while the third and fourth terms, giving nonlinear contributions to the equations of motion, correspond to the self-interaction of the field.

A situation richer in physical content arises when, instead of a real field, we have a complex field and the potential:

$$V(\varphi) = \frac{\mu^2}{2}\overset{*}{\varphi}\varphi + \frac{h^2}{4}(\overset{*}{\varphi}\varphi)^2.$$

Here instead of a discrete reflection symmetry we have a continuous symmetry under rotations of the complex φ plane. The potential energy minimum corresponds to $\overset{*}{\varphi}\varphi = |\varphi_0|^2 \equiv \mu^2/h^2$, which defines a circle in this plane. Therefore, the ground state is infinitely degenerate.

To study small deviations from equilibrium, we shift the field by an amount φ_0 corresponding to the ground state: $\varphi(x) \to u(x) = \varphi(x) - \varphi_0$. The choice of a specific value of φ_0, i.e., the choice of a point on the circle mentioned above, breaks the symmetry.

For definiteness, let φ_0 be real and positive. Then $u_1 \equiv \mathrm{Re}\, u = \mathrm{Re}\, \varphi - \varphi_0$, $u_2 \equiv \mathrm{Im}\, u = \mathrm{Im}\, \varphi$, and for V we obtain the expression $V = V(\varphi_0) + \mu^2\mu_1^2 + \mu h u_1(\overset{*}{u}u) + (h^2/4)(\overset{*}{u}u)^2$, from which we see that the real field u_1 has mass $\sqrt{2}\mu$ and the field u_2 is massless. The reason for this difference in the properties of the fields u_1 and u_2 is clearly seen from the form of the potential V (Fig. 19). Near the equilibrium position there is a potential barrier in the radial direction, but there is no such barrier in the tangential direction. In the vicinity of the value of φ_0 chosen above, the radial direction coincides with the direction of the real axis, and the tangential direction coincides with the imaginary axis.

The appearance of massless degrees of freedom always accompanies the spontaneous breakdown of any continuous symmetry, since such degrees of freedom are related to the infinitely degenerate vacuum state. In quantum field theory this means that massless scalar mesons necessarily appear when

Figure 19. Potential function of a model with spontaneous symmetry breaking in the case of a complex scalar field.

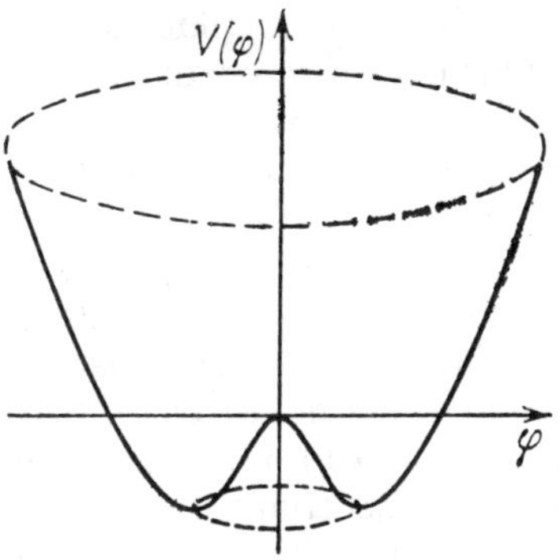

a continuous symmetry is spontaneously broken. This is referred to as the Goldstone theorem, and the massless particles are called "Goldstone bosons." The Goldstone theorem is the relativistic analog of the Bogolyubov theorem, discovered earlier in quantum statistics, which states that long-range forces appear in the spontaneous breakdown of a continuous symmetry.

The Higgs effect

A nontrivial application of spontaneous symmetry breaking arises when a scalar field interacts with other fields, in particular, with a gauge vector field. We recall that in spontaneous symmetry breaking it is the symmetry of the solution that is broken, not that of the Lagrangian. The shift of the scalar field by a constant as described above leads to the appearance of new structures in the Lagrangian. This fact can be used to introduce a gauge field mass without spoiling the gauge invariance of the Lagrangian. A massive vector field of this type cannot be free. It must interact with the scalar field, thereby serving as a means of breaking the symmetry spontaneously.

We know from the preceding section that, in addition to terms linear in the gauge field, the Lagrangian of the gauge interaction also contains a term of the form $|\varphi|^2A_\mu^2$. After a shift of the scalar field $\varphi'(x) = \varphi(x) - \varphi_0$ this term gives, along with the terms $|\varphi'|^2A_\mu^2$ and $|\varphi'|A_\mu^2$ describing the interaction between the fields $\varphi'(x)$ and $A_\mu(x)$, the term $\varphi_0^2A_\mu^2$, which gives the vector field a mass equal to $\sqrt{2}\mu/h$.

Above we saw that massless Goldstone bosons correspond to degrees of freedom associated with infinite degeneracy of the vacuum. In the spontaneous breakdown U(1) symmetry this is the degree of freedom associated with motion along the circle in the complex φ plane. In this direction the field functions change by phase transformations. It is easy to show that by a special choice of the phase the contribution of this degree of freedom can be cancelled at each point, i.e., the dependence of the Lagrangian on the corresponding field variable can be eliminated.

Therefore, in place of the original massless vector field and the complex scalar field with negative squared mass, we obtain a massive vector field and one massive real scalar field. The massless Goldstone boson has vanished as a result of the gauge transformation. This process of transforming a massless gauge field into a massive one with the simultaneous disappearance of the Goldstone boson is called the Higgs effect (or mechanism), and the massive scalar particle which appears is called a Higgs boson.

The Higgs effect leads to a redistribution of the degrees of freedom: one of the two real fields forming the complex scalar field is transformed into the third (longitudinal) component of the vector field. This redistribution is related to the fact that the massless vector field is transverse, i.e., it has two components, while the massive vector field has three components. Of course, the total number of physical degrees of freedom remains unchanged: $2+2 = 1+3$.

The Higgs effect is important in particle theory for the following reason. The theory of the interaction of a massive vector field with other fields constructed in the standard manner is not renormalizable, because the propagator of a massive vector field does not vanish in the ultraviolet limit [it goes as $m^{-2}k_\mu k_\nu/(k^2 - m^2)$]. However, theories based on gauge invariance and containing massless vector fields are renormalizable. The propagators of massless vector fields fall off at large k as $(1/k^2)$, and the requirement of invariance imposes rigorous constraints on the possible structures of divergent radiative corrections, allowing only a finite number of them.

The gauge invariance of the Lagrangian is preserved in spontaneous symmetry breaking. This forms the basis of the proof that the results of calculating the elements of the scattering matrix are independent of the choice of gauge. In particular, it is possible to choose a gauge in which the Goldstone bosons are absent and the vector field has a mass. Another possible choice of gauge is that in which the vector field propagator is transverse $[1/(k^2 - m^2)][g_{\mu\nu} - k_\mu k_\nu/k^2]$. The first choice is useful because in it the S-matrix is explicitly unitary in the space of physical states, while in the second gauge it is easier to prove renormalizability. In fact, the transverse propagator falls off with increasing k as $1/k^2$. Moreover, it can be shown that all the constraints on the structure of the possible divergences remain valid. The contributions to the scattering matrix from the unphysical states which are present in this gauge—fictitious Goldstone bosons, extraneous polarization states of the vector field, and the Faddeev–Popov ghost fields corresponding to propagators with poles at $k^2 = 0$—cancel in the sum.

Therefore, quantum field models in which the vector fields obtain a mass via spontaneous symmetry breaking are renormalizable. This remarkable fact permitted the construction of a renormalizable theory of weak interactions, which will be the topic of the following chapter.

Chapter 5
Theory of the electroweak interaction

16. The current status of quantum electrodynamics

The electron anomalous magnetic moment

As is well known, the Dirac equation for an electron interacting with an electromagnetic field contains a term corresponding to the magnetic interaction. The strength of the latter is described by the electron magnetic moment μ, which is $\mu_0 = e\hbar/2mc$, the Bohr magneton. Vacuum polarization effects lead to an additional contribution referred to as the anomalous magnetic moment. The simplest correction to μ comes from the one-loop vertex diagram shown in Fig. 20. It was calculated by Schwinger in 1948 and is

$$\delta\mu = (\alpha/2\pi)\mu_0 \simeq 0.00116\mu_0.$$

This deviation from the Bohr magneton was observed at almost the same time in experiment: the measured magnetic moment proved to be close to the Schwinger value.

Later, both the theoretical and the experimental values were refined many times. The present situation is the following. The anomalous magnetic moment has been calculated through the four-loop approximation. The result can be written as

$$\frac{\mu_{\text{theor}}}{\mu_0} = 1 + \frac{1}{2}\left(\frac{\alpha}{\pi}\right) + C_2\left(\frac{\alpha}{\pi}\right)^2 + C_3\left(\frac{\alpha}{\pi}\right)^3 + C_4\left(\frac{\alpha}{\pi}\right)^4,$$

where $C_2 = -0.328\,478...$, $C_3 = 1.1765(13)$, and $C_4 = -0.8 \pm 2.5$.

The numbers in parentheses are the significant figures of the uncertainty to one standard deviation, for example, $1.1765(13) = 1.1765 \pm 0.0013$. The coefficient C_2 has been calculated analytically; it involves the computation of five diagrams. The three-loop contribution C_3 involves 72 diagrams; it has been possible to calculate 51 of these exactly using a computer for the analytical calculations. The error in C_3 (and in C_4) comes from the approximate numerical calculation of the remaining diagrams (in the case of C_4 all the 891 four-loop diagrams have to be done in this manner).

Comparison of the theoretical result to the experimental value:

$$\mu_{\text{exp}}/\mu_0 = 1.001\,159\,652\,200(40),$$

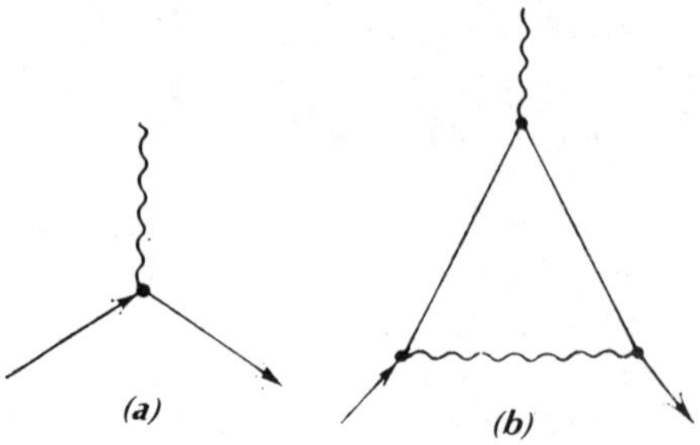

Figure 20. Feynman diagrams contributing to the electron magnetic moment: (a) the Bohr magneton, (b) the Schwinger correction.

requires use of the numerical value of the fine structure constant. Taking its most accurate value $\alpha^{-1} = 137.035\,963(15)$, obtained using the Josephson effect, we find $\mu_{\text{theor}}/\mu_0 = 1.001\,159\,652\,460(147)$, where most of the error is due to the uncertainty in α.

We see that the theoretical and experimental values agree to within 2 parts in ten billion, which is a record in physics.

Other effects

This particular example is very impressive. In the case of other quantum electrodynamical effects, either the experimental accuracy is not very good or the theoretical results are constrained by the need to include corrections arising from sources outside electrodynamics.

For example, the experimental value of the hyperfine splitting of the $1s_{1/2}$ level in the hydrogen atom due to the interaction of the electron with the magnetic field of the proton is presently known to 13 decimal places, while the theoretical value is known only to 7, even when the finite size of the proton is taken into account.

The Lamb shift, i.e., the fine splitting of the $2s_{1/2}$ and $1p_{1/2}$ levels in hydrogen is known with an error of order 10^{-6}:

$$s_{\text{exp}} = s_{\text{theor}} = 1057.89(2)\ \text{MHz}.$$

Here effects arising from sources outside quantum electrodynamics contribute at the level 10^{-5} and are included in the theoretical value.

On the whole, it can be stated that all the experimental data on a large number of phenomena are in complete quantitative agreement with the results calculated using perturbative quantum electrodynamics whenever the strong interaction is unimportant or can be taken into account.

Conclusion

Returning to the electron anomalous magnetic moment, we stress the fact that the level of agreement between theory and experiment in that case is of fundamental significance for local quantum field theory in general. It indicates that all the basic postulates of this theory are true to a high degree of accuracy. This pertains, in particular, to the crucial property of locality, which is intimately related to the origin of divergences and the need to renormalize.

Beginning in the 1940s and continuing until the present, attempts have been made, on the basis of the work of Markov, Tamm, and D.I. Blokhintsev, to construct a systematic nonlocal variant of quantum field theory free from the difficulties associated with ultraviolet divergences. The nonlocal modification unavoidably involves the introduction of a new fundamental constant into the theory: the "fundamental length" L characterizing the short-distance scale at which effects of nonlocality begin to manifest themselves. Without going into the details of the significant progress which has been made in the construction of a nonlocal theory, we note that there are no experimental indications of the existence of nonlocal effects. The results presented above on the electron anomalous magnetic moment and also those on certain other effects lead to the conclusion that $L < 10^{-16}$ cm.

17. The weak interaction with an intermediate boson

Currents and the Fermi interaction

The generalization of the quantum field model for neutron β decay proposed by Fermi to other processes has allowed the construction of a universal theory of weak interactions which gives a good description of a large set of experimental data. In this theory the elementary weak interaction event is associated with the product of four spinor field operators taken at the same point, with the dimensionful coupling constant $G_F \approx 10^{-5}\,\mathrm{GeV}^{-2}$. Here it turns out that the interaction strength, which depends on the process, is characterized by the dimensionless product $G_F E^2$, where E is the typical energy (mass) of the process. It is easy to see that this quantity is considerably smaller than unity up to energies of order 100 GeV.

Before describing the structure of the weak interaction in more detail, let us briefly explain the terminology. An expression of the form $j(x) = \bar{u}(x) O v(x)$, which is bilinear in the fermion field operators $u(x)$ and $v(x)$ and where O is some combination of matrices including the Dirac matrices, is called a current. In the special case where u and v correspond to the same field and the matrices O are the Dirac matrices γ_μ, this expression is the usual electromagnetic current appearing in the interaction Lagrangian of quantum electrodynamics, $e\bar{u}(x)\gamma_\mu u(x) A_\mu(x)$. Depending on the form of the matrix O, the current $j(x)$ can be a scalar ($O = 1$), a pseudoscalar ($O = \gamma_5$), a vector ($O = \gamma_\mu$), an axial vector ($O = \gamma_\mu \gamma_5$), or a tensor ($O = \gamma_\mu \gamma_\nu$) under Lorentz transformations. In the late 1950s, after the discovery of parity violation, it was found that the weak current is the sum of a vector and an axial vector, i.e., $j(x) = j_V(x) + j_A(x)$.

The four-fermion Lagrangian of the weak interaction is written as the product of two currents:

$$\frac{G}{\sqrt{2}} j^+(x)\, j(x) = \frac{G}{\sqrt{2}} (j_V^+ j_V + j_A^+ j_A + j_A^+ j_A + j_A^+ j_V).$$

Here the terms corresponding to the product of a vector and an axial vector lead to transitions in which parity is not conserved. (We recall that vectors and axial vectors behave identically under all transformations except spatial reflections, under which they differ by a sign.)

The weak current is split into terms containing only lepton fields and terms containing only hadron fields, $j(x) = j^l + j^h$. These are called leptonic and hadronic weak currents. Products of two leptonic currents describe purely leptonic weak processes, for example,

$$\mu^- \to e^- + \bar{\nu}_e + \nu_\mu, \quad \nu_\mu + e^- \to \nu_\mu + e^-, \text{etc.}$$

Products of leptonic and hadronic currents correspond to so-called semileptonic processes, which involve both leptons and hadrons. Examples of such processes are

$$n \to p + e^- + \bar{\nu}_e, \quad \Lambda \to p + e^- + \bar{\nu}_e, \quad \text{and so on.}$$

Finally, nonleptonic weak processes involve only hadrons. Strangeness-changing nonleptonic interactions are manifested in the purely hadronic decays of hyperons and K mesons. Strangeness-conserving processes lead to so-called p-odd nuclear forces. These forces, predicted theoretically, were discovered in 1964 by Yu.G. Abov and P.A. Krupchitskiĭ and studied quantitatively in 1967 by V.M. Lobashov and V.A. Nazarenko.

After the quark structure of hadrons was established, a change occurred in the representation of the structure of the weak current. Instead of hadron currents, currents constructed from quark fields were used.

When the fermion fields u and v correspond to particles with different values of the electric charge, the current $j = \bar{u}Ov$ is called a charged current.

If the electric charges of the fields u and v are the same, the current is termed neutral.

We note that in this terminology the ordinary electromagnetic current is a neutral current. Neutrality of the electromagnetic current means that the charge of a particle is not changed in an interaction, since the carrier of the interaction (the photon) is neutral.

The weak processes first discovered experimentally were those associated with charged currents. Examples of such processes are

$$n \to p + e^- + \bar{\nu}_e, \quad \mu^- \to e^- + \bar{\nu}_e + \nu_\mu,$$

$$\nu_\mu + n \to p + \mu^-, \quad \Lambda \to p + e^- + \bar{\nu}_e,$$

and there are many others. The existence of weak processes due to neutral currents was predicted theoretically by the unified theory of the weak and electromagnetic interactions constructed in the 1960s. Several years later these currents were discovered experimentally. In 1973 experiments carried

Figure 21. Illustration of the transition from the four-fermion interaction to the interaction transmitted by an intermediate boson.

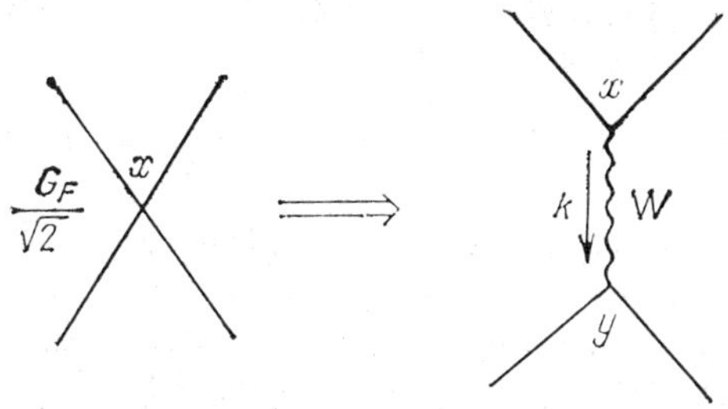

out at the European Center for Nuclear Research (CERN) on the scattering of muon neutrinos on nucleons discovered the muonless processes:

$$\nu_\mu + N \rightarrow \nu_\mu + \text{hadrons}, \quad \bar{\nu}_\mu + N \rightarrow \bar{\nu}_\mu + \text{hadrons}.$$

It was then concluded that there exists a weak interaction described by the product of a neutral neutrino current with neutral quark currents. In 1978, while studying the optical activity of vapors of atomic bismuth, L.M. Barkov and M.S. Zolotorev experimentally discovered the weak neutral electron current. Finally, the other weak neutral leptonic currents were discovered in 1982.

The intermediate boson

As already mentioned in the second chapter, the four-fermion model successfully describes a number of regularities observed experimentally in weak interaction processes.

However, since the coupling constant has dimensions, the Fermi model is not renormalizable and thus has no hope of being a consistent theory. The power law growth of the matrix elements with energy contradicts unitarity, so the expressions obtained in lowest order perturbation theory are not applicable for describing phenomena at high energies (on the order of hundreds of GeV). The boundary of the region where the four-fermion theory is applicable is referred to as the "unitarity limit." Its existence was first predicted by Blokhintsev.

Meanwhile, the low energy predictions are not changed by assuming that the (current $\times$ current) form of the four-fermion weak interaction Lagrangian is not fundamental, but instead arises because weak currents interact with each other via some massive intermediate field. In other words, these predictions are unchanged if it is assumed that the interaction of two currents is not local, but is spread over different spacetime points, as shown in Fig. 21.

As an analogy we can take a quantum field model similar to quantum electrodynamics, but in which the "photons" have a quite large mass and are described by a vector field B_μ. If we restrict ourselves to processes in which the heavy "photons" appear only in virtual states, and are absent from the initial and final states, such processes can be described by formally eliminating the vector fields B_μ from the theory and replacing the interaction Lagrangian $ej_\mu(x)B_\mu(x)$ by some effective nonlocal Lagrangian $e^2 j_\mu(x)D_{\mu\nu}(x-y)j_\nu(x)$, where $D_\mu(x-y)$ is the "photon" propagator. As already discussed in connection with the Yukawa model, when the interaction carrier has a nonzero mass, the interaction range is finite.

It can therefore be concluded that the fundamental Lagrangian of the weak interaction has a form similar to the Lagrangian of electrodynamics, $g j_\mu(x) W_\mu(x)$, where g is a dimensionless coupling constant and W_μ is the vector field transmitting the interaction. The corresponding particle is called an "intermediate vector boson." With this hypothesis it is possible to explain why the four-fermion theory gives a good description of certain processes, and its range of applicability can be determined. In fact, if the mass of the intermediate boson is sufficiently large, and the momentum transfers and all energies are much smaller than M_W, then intermediate bosons can arise only in virtual states. Here the interaction resulting from intermediate boson exchange is almost indistinguishable from the four-fermion interaction, since the propagator of the field W_μ in the momentum representation $g^{\mu\nu}/(M^2-k^2)$ in this case is approximately equal to the constant $g^{\mu\nu}/M^2$. In the coordinate representation this expression corresponds to a δ function, $g^{\mu\nu}/M^2\delta(x-y)$, so the points x and y practically coincide.

Therefore, the two-current interaction in the low energy approximation can effectively be described by the local expression $g j_\mu(x) W_\mu(x)$. It is easy to see that the Fermi constant G_F has a simple relation to the dimensionless constant g:

$$G_F/\sqrt{2} = g^2/8M_W^2. \qquad (17.1)$$

This expression is easily derived by comparing the diagrams given in Fig. 21 and neglecting the momentum k of the intermediate boson compared to its mass M_W.

The problem of renormalizability

The fact that the coupling constant is dimensionless makes us optimistic about the possibility of constructing a renormalizable model of weak interactions.

However, here the situation is completely different from that in the known renormalizable theories such as quantum electrodynamics or Yang–Mills field theory. Let us briefly recall the problem. It has already been discussed in detail at the end of Sec. 15. In contrast to massless vector fields, a massive vector field has a propagator which contains, in addition to the transverse part, a longitudinal piece determining its behavior at large momenta and leading to ultraviolet divergences which are stronger than those for the transverse propagator. Therefore, theories with massive vector fields are nonrenormalizable. The only possible way of constructing a renormalizable quantum field model with a massive vector field is based on use of a spontaneous symmetry breaking mechanism.

18. The Glashow–Weinberg–Salam model

The choice of gauge group

For quite a long time, quantum electrodynamics was the only consistent theory in particle physics which gave quantitative predictions in agreement with experiment. It was also the first quantum field model possessing gauge invariance. The principle of local gauge invariance can also serve as the basis for a theory of the weak interaction. Here the mass of the interaction carriers appears as a result of spontaneous symmetry breaking. The general principle on which the models of the weak and electromagnetic interactions are based allows the formal unification of these models into a single gauge theory.

The gauge group of the unified theory must have a dimension of at least three. This is because there must be at least three vector fields in the theory—the electromagnetic field and two fields corresponding to the charged intermediate bosons $W^{\pm}$. (We recall that when this theory was constructed the existence of neutral currents was not yet known. However, as we shall see, in the constructed model they necessarily accompany the charged currents.)

As explained in Sec. 14, the number of gauge fields coincides with the dimension of the group. The simplest unitary group of dimension three is SU(2). However, when this group is used it is not possible to obtain parity violation in the weak interaction and its conservation in electromagnetism. In fact, the weak charged current (for simplicity, we shall first consider only leptons, and discuss the introduction and role of quark currents at the end of this section) can be written as

$$j = \bar{\nu}_e \gamma_\mu (1+\gamma_5) e^- = \bar{\psi}_e \gamma_\mu (1+\gamma_5)\tau^+ \psi_e,$$

where $\psi_e = \binom{\nu_e}{e^-}$ is a doublet containing the electron and neutrino field functions and $\tau^+ = \begin{pmatrix} 0 & 1 \\ 0 & 0 \end{pmatrix}$.

The arguments for other leptons are similar. The matrix $(1+\gamma_5)\tau^+$ and the matrix $(1+\gamma_5)\tau^-$ corresponding to the conjugate current are generators of SU(2). The third generator of this group is the matrix $(1+\gamma_5)\tau_3$, which corresponds to a neutral, parity-violating current. We note that the common factor $(1+\gamma_5)$ has no effect on the group structure. The latter is determined by the τ matrices, which, as is well known, are the generators of SU(2). There is no place in this scheme for a neutral, parity-conserving current which could play the role of the electromagnetic current. Therefore, to include the electromagnetic interaction in this scheme it is necessary to choose a group of dimension four. S. Glashow showed that a possible choice is the group SU(2)×U(1). This group served as the basis for the unified model of the electromagnetic and weak interactions of leptons proposed in 1967 by Weinberg and Salam.

The physical structure

Let us now briefly describe this model. Since it is quite complicated, particularly where fermions are involved, we shall restrict ourselves mainly to a description of the vector fields transmitting the weak and electromagnetic interactions. Local gauge invariance under the group SU(2)×U(1) requires the existence of two vector gauge fields—a three-component field $V_\mu^a(x)$ ($a = 1,2,3$) and a one-component field $B_\mu(x)$, whose interactions with matter fields are characterized by the two coupling constants g_1 and g_2. The fields $V_\mu^1(x)$ and $V_\mu^2(x)$ are charged and the fields $V_\mu^3(x)$ and $B_\mu(x)$ are neutral.

Spontaneous symmetry breaking leads to masses for the fields $V_\mu^1(x)$ and $V_\mu^2(x)$ and for a superposition of the fields $V_\mu^3(x)$ and $B_\mu(x)$: $Z_\mu = V_\mu^3 \cos\theta_W - B_\mu \sin\theta_W$. The parameter θ_W, called the Weinberg angle, has a simple expression in terms of the coupling constants: $\theta_W = \arcsin[g_1(g_1^2 + g_2^2)^{-1/2}]$. The combination $A_\mu = B_\mu \cos\theta_W + V_\mu^3 \sin\theta_W$, which is orthogonal to Z_μ, remains massless and interacts with the usual electromagnetic current. It can therefore be identified as the photon field. The massive vector fields $W_\mu^\pm = V_\mu^1 \pm iV_\mu^2$ and Z_μ, respectively, interact with the charged and neutral parity-violating currents.

The use of the Higgs mechanism to give the vector fields masses unavoidably leads to the appearance of a massive scalar particle in the model—the Higgs boson. Up to now it has not been seen experimentally. One reason for this might be that it has a large mass inaccessible at present-day accelerators.

Without going into details about the fermions, we only note that each pair of leptons (e^-, ν_e), (μ^-, ν_μ), and (τ^-, ν_τ) is introduced into the model independently.

Comparison with experiment

Let us now consider the main predictions of the model and their experimental status. One of the most important is the prediction that neutral weak currents exist. For several years after the construction of the model this prediction was viewed as a defect. As we noted in the preceding section, neutral current processes were discovered experimentally in 1973. This was the first triumph of the Weinberg–Salam model.

The most direct test of the theory was the search for the intermediate bosons. The order of magnitude of their mass can be estimated from Eq. (17.1), taking $g^2 \sim e^2 \sim 1/137$. This gives a value of several tens of GeV. The Weinberg–Salam model gives more accurate predictions. When radiative corrections, first calculated by D.Yu. Bardin and M. Veltman, are included the predicted values are $M_W = 82.5 \pm 2.5$ GeV and $M_Z = 93.3 \pm 2.5$ GeV. Until recently it was not possible to reach such high energies at accelerators. The intermediate vector bosons therefore remained undiscovered for a long time. The creation of these particles was observed only in 1983 in special experiments carried out at the CERN $p\bar{p}$ collider, and their masses were measured. The improved experimental values are $M_W^{\rm exp} = 80.0 + 0.7$ GeV and $M_Z^{\rm exp} = 91.2 + 0.2$ GeV. The latter one was obtained in 1989 at LEP and SLC e^+e^- colliders, where also the width of Z-boson was first measured. These results are the value of the Weinberg angle ($\sin^2\theta_W = 0.230 + 0.005$) measured in several different experiments are in good agreement with the theoretical predictions.

We have presented the most striking examples of the experimental confirmation of the Weinberg–Salam model. We again stress the fact that this model fully reproduces the low energy results of the universal four-fermion theory and, in contrast to the latter, is renormalizable.

Anomalies and the quark extension

This model has an important feature associated with parity violation. Since the theory contains triangle diagrams in which all three internal lines are fermions and of the external lines two are vectors and one is an axial vector, the so-called axial anomaly arises.

An anomaly is usually interpreted as the violation of some conservation law, following from the symmetry properties of the Lagrangian of the classical theory, by the quantum corrections calculated using this Lagrangian. A characteristic feature of quantum anomalies is the impossibility of eliminating them by a redefinition (renormalization) of certain quantities or parameters.

In the case in question here, the contribution of the triangle diagram makes it impossible to simultaneously satisfy the conditions of vector current conservation and partial conservation of the axial current. A consequence of this is that the usual ultraviolet renormalizations of the vector vertex containing the triangle anomal diagram as a subdiagram do not remove all divergences. This has catastrophic consequences for the renormalizability of the entire theory.

The only possible salvation is that the contributions of different fermions, whose propagation can be associated with the internal lines of a given triangle diagram, cancel in the sum. It can be shown that the contribution of each fermion to the axial anomaly is proportional to its electric charge. The condition for cancellation of the axial anomalies is therefore that the sum of the charges of all the fermions is equal to zero.

This cancellation can be obtained by taking into consideration the quark weak interaction, the effect of which is manifested experimentally as the hadron weak interaction. The quark weak currents have been constructed in studies by Glashow and others. The unified theory of electroweak interactions of leptons and quarks is called the Glashow–Weinberg–Salam model.

If each pair of leptons is associated with a pair of quarks, the anomaly cancellation arises naturally. For example, if we associate the electron and electron neutrino with the u and d quarks, we find that the total electric charge is zero:

$$\sum Q = -1 + 0 + 3(2/3 - 1/3) = 0. \tag{18.1}$$

The factor of 3 multiplying the quark charges corresponds to the three possible quark colors.

Generations

The sets of leptons and quarks formed in this manner are called fermion generations. The first generation contains the lightest leptons e^-, ν_e and the lightest quarks u and d.

The second generation contains the muon and its neutrino, μ^- and ν_μ, and the c and s quarks. Here again the condition (18.1) is satisfied. The crucial factor for this is the existence of the fourth quark c, which was theoretically predicted on the basis of quark–lepton symmetry considerations long before its experimental discovery in 1974.

The third generation contains the τ lepton, discovered in 1975, and its neutrino, and also the b and t quarks. The b quark was discovered experimentally in 1987. Up to now (Nov. 1989) there is no experimental indication to the existence of the t quark. The experimental lower bound for its mass is 70 GeV. And the theoretical estimates for the upper band give $m_t < 180$ GeV.

The fermion masses increase significantly in going from one generation to the next. For example, $m_\mu \approx 200\ m_e$ and $m_\tau \approx 20\ m_\mu$. The quark masses increase similarly.

At the present time the reasons for the grouping of fermions into generations are not understood at all. The total number of generations is also unknown. (However, there is indirect evidence from astrophysical observations indicating that the number of light neutrino types is equal to three or four.) From the measurements of the width of Z-boson it was obtained that the number of light neutrinos is equal to 3.2 to 0.2. This result also indicates the feasible number of generations.

This large number of known quarks and leptons suggests that they themselves might be the bound states of some more elementary objects. A number of such hypotheses have been made, but there is nothing even close to a systematic theoretical model, nor any experimental evidence.

Summarizing our discussion of the Glashow–Weinberg–Salam model, we again emphasize the fact that it, at present, is universally recognized as the model of the electromagnetic and weak interactions which is consistent with the experimental data. However, so far no Higgs bosons have been discovered, so there is not yet any experimental confirmation of the spontaneous symmetry breaking and mass generation mechanism used in the model. Without going into details, we also note that the nonzero vacuum expectation value of the scalar field in this model leads to masses not only for the gauge vector fields, but also (owing to special interactions of the Yukawa type) for the fermions. In order to account for the differences between the fermion and intermediate boson masses, it is necessary to assume that the Yukawa constants are quite small and, moreover, that they are different for each type of fermions. The presence of extra, nongauge interactions characterized by different coupling constants considerably spoils the simplicity and economy of the model.

Chapter 6
Quantum chromodynamics

19. The composite structure of hadrons

Deep inelastic scattering of leptons on hadrons

As mentioned in Sec. 8, the study of the high energy scattering of leptons on nucleons led to the discovery that nucleons have a complex structure. The scattering of leptons on hadrons involves transfer of energy and momentum via a virtual particle: a photon when the leptons are electrons or muons, and also an intermediate boson when the leptons are neutrinos. For brevity, we shall speak of only the virtual photon, with the understanding that intermediate boson exchange may also be possible.

In Fig. 22 we schematically show the process of deep inelastic scattering of a lepton (e, μ, or ν) on a nucleon, in which the nucleon N is transformed into a system of hadrons X. The kinematical conditions $(p+q)^2 \gg M_N^2$ and $-q^2 \gg M_N^2$, respectively, ensure that a large number of hadrons can be created and that the virtual photon can probe the nucleon to distances much smaller than the nucleon size.

As we pointed out in Sec. 8, inclusive characteristics are technically the simplest to study and give a large amount of information on the hadron structure. By "inclusive characteristics" we mean the characteristics of processes for which only the scattered lepton is detected in the final state, while the final hadron system can be anything.

The parton model

The available experimental data are described quite well by the parton model. In this model the virtual photon interacts incoherently with pointlike constituents of the hadron—the so-called partons. By "incoherent" we mean that the probability for the photon to interact with a set of partons is equal to the sum of the probabilities for interacting with each one separately.

In the parton model it is assumed that a deep inelastic scattering process occurs in two independent stages, as shown in Fig. 23. In the first stage the virtual photon interacts with one of the partons, which acquires a large recoil momentum. In the second stage "hadronization" occurs: the partons are transformed into hadron jets which are observed experimentally.

The experimentally measured dependence of the inclusive cross sections on the square of the momentum transfer q^2 and the energy transferred to the hadron system is in qualitative agreement with the representation of partons as pointlike objects. It also is consistent with the interaction between partons being quite weak at large $|q^2|$ ($|q^2| \gg 1$ GeV^2).

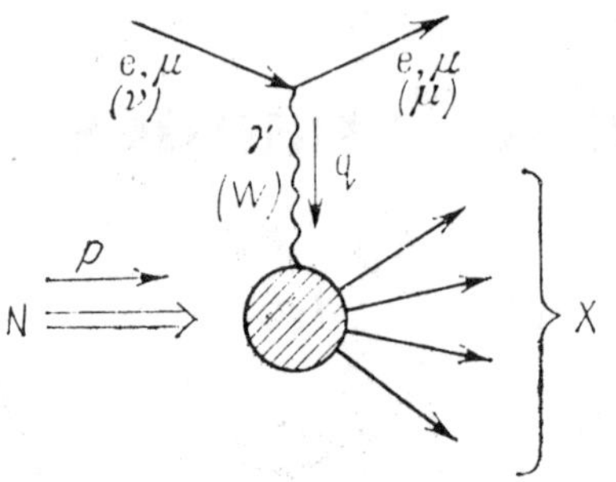

Figure 22. Schematic depiction of the deep inelastic scattering of a lepton on a nucleon in which the nucleon is transformed into a system of hadrons.

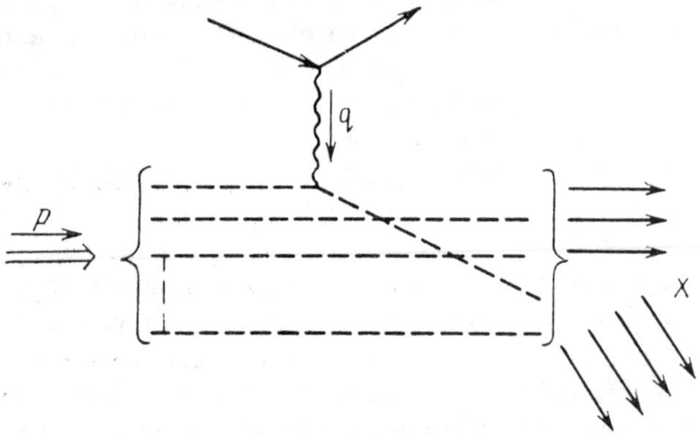

Figure 23. Representation of a deep inelastic scattering process in the parton model.

Quarks and gluons

Analysis of the data on the deep inelastic scattering of electrons, muons, and neutrinos also leads to the conclusion that charged partons can be identified with quarks. However, not all the partons correspond to quarks. Roughly half the momentum of a nucleon is carried by electrically neutral objects called "gluons." By assumption, they transmit the interaction between quarks and, in particular, "glue" them together into hadrons.

The exchange of gluons between quarks must lead to an interaction which is weak at short distances ($<10^{-15}$ cm) and strong enough at large distances ($\gtrsim 10^{-14}$ cm) to ensure that hadrons are stable.

The quark–gluon interaction and asymptotic freedom

We have seen that in quantum field theory the strength of the interaction is determined by the value of the effective coupling constant. As discussed above in Sec. 13, the behavior of the effective coupling constant required for the quark interaction occurs in non-Abelian gauge theories. In these theories the interaction strength grows with distance, and as the distance decreases it tends to zero as in expression (13.7):

Figure 24. Schematic depiction of the nucleon structure. The open circles represent valence quarks, and the darkened ones represent sea quarks. The wavy lines depict gluon exchanges.

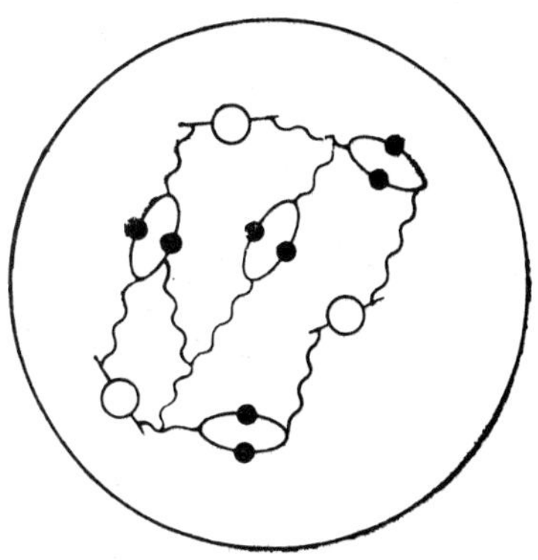

$$\alpha_s(r) \sim \left(c \ln \frac{1}{r\Lambda}\right)^{-1}$$

This automatic switching off of the interaction at short distances is called "asymptotic freedom."

We therefore obtain the following picture of hadron structure. Depending on its quantum numbers, a hadron contains a certain minimum number of quarks and antiquarks, sometimes referred to as "valence" quarks. For example, baryons contain three valence quarks and mesons contain a valence quark and antiquark. The valence quarks interact with each other by exchanging virtual gluons—quanta of the Yang–Mills field. In turn, the gluons create virtual quark–antiquark pairs, which are called "sea" quarks to distinguish them from valence quarks. In addition to all this, there are processes in which virtual gluons are emitted by sea quarks and by the gluons themselves, and so on.

Figure 24 schematically depicts the structure of a nucleon. The open circles correspond to valence quarks, and the darkened ones to sea quarks. It should be noted that either of these can be located at any point inside the hadron. Actually, this picture is merely suggestive—it would be more correct to say that a hadron corresponds to an infinite set of such pictures with different numbers of virtual quarks and gluons.

20. The gauge theory of the quark–gluon interaction

The quark color symmetry

The discovery of the asymptotic freedom of non-Abelian gauge theories opened up the possibility of constructing a quantum field model of quark interactions which would describe the experimental data on deep inelastic processes. This model is based on the quark color symmetry group $SU_c(3)$ with which we became acquainted in Sec. 8. Although it is impossible to observe

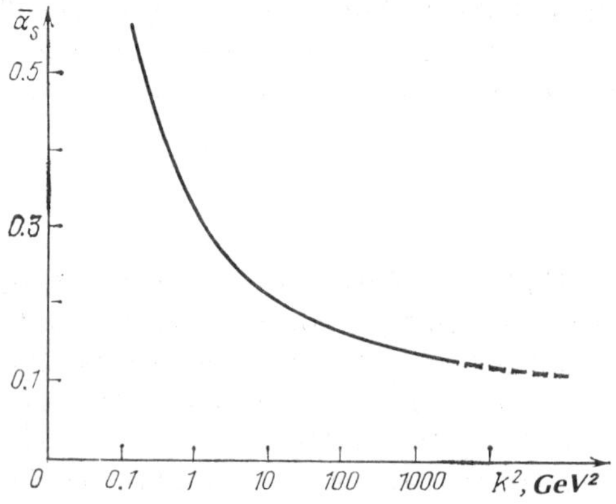

Figure 29. Behavior of the effective coupling α_s in quantum chromodynamics as a function of the squared momentum transfer.

$$\bar{\alpha}_s^{(2)}(k^2) = \frac{\alpha_s}{1 + \beta_1 \alpha_s l + (\beta_2/\beta_1)\alpha_s \ln(1 + \beta_1 \alpha_s l)}, \tag{20.2}$$

where

$$l = \ln \frac{k^2}{\mu^2}, \quad \alpha_s = \bar{\alpha}_s(\mu^2),$$

$$\beta_1 = (33 - 2n)/12\pi, \quad \beta_2(306 - 38n)/48\pi^2.$$

Changing over to the scale parameter Λ introduced above, we can write Eq. (20.2) approximately as

$$\bar{\alpha}_s^{(2)}(k^2) = \frac{1}{\beta_1 L} - \frac{\beta_2}{\beta_1^3}\frac{\ln L}{L^2}, \quad L \equiv \ln \frac{k^2}{\Lambda^2}. \tag{20.3}$$

The numerical value of Λ determined from the experimental data is of order 100 MeV. Here it is important to note that when using Eqs. (20.2) and (20.3) in the heavy quark mass range, n should be understood as the effective number of flavors, i.e., the number of quarks for which the relation $m^2 \ll k^2$ is satisfied and for which the logarithmic asymptote can be used to include them in intermediate states. In the range 1 GeV$^2 < k^2 <$ 100 GeV^2 we must take $n = 3$.

Under these conditions the second term on the right-hand side of Eq. (20.3), i.e., the two-loop contribution, is about 1/5 of the first term, and in this range $\bar{\alpha}_s$ itself is much smaller than unity. A semiquantitative graph of the dependence of $\bar{\alpha}_s$ on k^2 is shown in Fig. 29. From it we see that for $k^2 > 100$ GeV, $\bar{\alpha}_s$ falls off weakly with increasing k^2, varying in the range 1/7–1/10. More careful consideration shows that in this range it is possible to use two-loop perturbation theory improved by means of the renormalization group. On the other hand, this graph shows that $\bar{\alpha}_s$ grows rapidly with decreasing k^2 beginning at ~10 GeV^2, and for $k^2 < 1$ GeV^2 we enter the strong coupling region.

Therefore, the two-loop expression (20.2) reproduces not only the phenomenon of asymptotic freedom at large k^2, i.e., short distances, but also the rapid approach (with increasing distance) to the "hadronic physics regime," where strong coupling rules. The physical interpretation of the scale parameter Λ is that its numerical value represents the scale characteristic of strong coupling.

The experimental status of QCD

It follows from the above discussion that at sufficiently large momentum transfers QCD perturbation theory can be used to obtain numerical results for strong interaction processes. This procedure is complicated because the calculations involve quark–gluon subprocesses, while the characteristics which are observed pertain to hadronic processes. It is therefore necessary to solve the theoretical problem of finding the relation between the matrix elements for hadronic processes and those for the quark–gluon subprocesses.

In some cases, in particular, for so-called hard processes, it is possible to distinguish characteristics in whose calculation the contributions from large and small momentum transfers (small and large distances, respectively) can be separated. The dependence of the quantity of interest M on the small (p^2) and large (Q^2) invariant momentum variables is written in factorized form $M(Q^2,p^2) = \mathscr{M}(Q^2) \times F(p^2)$. Here the function $\mathscr{M}(Q^2)$ calculated using perturbative QCD describes the quark–gluon subprocess occurring at short distances, and the function $F(p^2)$ contains all the necessary information about the dynamics at large distances. Phenomenological functions $F(p^2)$ which are identical for different processes characterize the parton distribution in hadrons and the probabilities for hadron formation from parton configurations. The form of the functions F is determined from the experimental data on the simplest hard processes. As a rule, lepton–hadron processes are used for this.

Therefore, quantum chromodynamics as a theory is not as straightforward from the physical point of view as quantum electrodynamics. In fact, the quanta of the constituent fields—quarks and gluons—are not directly observable, like the quantum number color on which the gauge mechanism for the interaction of these fields is based. Nevertheless, there are convincing experimental arguments in favor of QCD. The most important of these are:

(1) The nature of the dependence of the inclusive cross sections for deep inelastic lepton–hadron scattering on the square of the momentum transfer q^2 predicted by QCD agrees with the experimental data.

(2) In electron–positron annihilation, events are observed in which three hadron jets are created. The dominant contribution to the e^+e^- annihilation cross section comes from the two-jet processes shown in Fig. 26. However, in addition to these, QCD predicts a sizable probability for the creation of hadron jets by gluons emitted by one of the quarks (Fig. 30). The angular and energy distributions of hadrons in the observed three-jet events are consistent with a gluon spin of 1. These results are viewed as experimental proof of the existence of gluons.

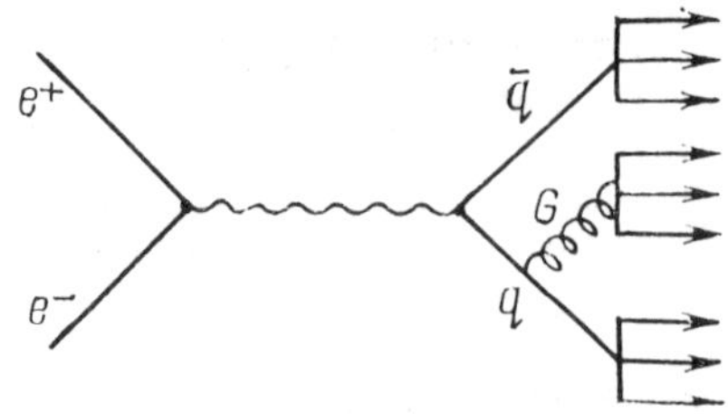

Figure 30. Diagram describing the creation of three hadron jets in e^+e^- annihilation.

The situation where the existence of a particle is deduced on the basis of the observation of its decay products is not novel in particle physics. For example, resonances are not observed directly; only the hadrons formed as a result of their decay are.

(3) QCD successfully describes the observed spectroscopy of heavy quarkonia—$c\bar{c}$ and $b\bar{b}$ bound states, and predicts the three-gluon decays of the J/ψ and γ mesons seen experimentally.

(4) The regularities observed in experiments on the production of hadrons with large transverse momentum in hadron collisions arise naturally in QCD as a result of the direct gluon–gluon interaction.

(5) The reliability of QCD is attested to by the fact that the values of the parameter Λ, a fundamental quantitative feature of QCD, determined experimentally in various processes are very close to each other, lying in the range 100–200 MeV.

Therefore, even though the fundamental physical concepts on which QCD is based are not accessible to direct observation, the excellent qualitative and good quantitative agreement between its predictions and experiment indicate that this quantum field model is a realistic theory of the strong interaction.

In recent years, quark–gluon concepts have also been used in nuclear physics to describe collisions of hadrons and nuclei with nuclei involving large momentum transfers (on the order of 1 GeV per nucleon and larger). This line of research, referred to as relativistic nuclear physics, originated in the studies of Baldin. This is a new application of QCD which, in turn, serves as a source of additional information about the details of the quark–gluon interaction.

The interaction at large distances

Of course, to be regarded as the theory of the strong interaction, QCD must describe all of hadron physics. Here one unavoidably encounters the problem of quantitatively describing phenomena occurring at large ($\gtrsim 10^{14}$ cm) distances.

The formal extension of the expression for the effective charge (20.1) [or (20.2)] to small k^2 (large distances) predicts that interactions grow stronger with increasing distance. A rigorous quantitative analysis using perturbation theory obviously cannot be carried out in this region owing to the large

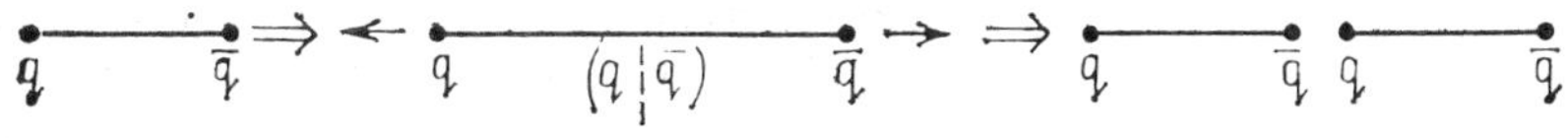

Figure 31. Schematic depiction of the formation of new colorless states as a quark and antiquark are pulled apart.

value of the expansion parameter. Nevertheless, the strengthening of the (attractive) interaction at large distances qualitatively explains the binding of quarks and gluons into hadrons.

Numerical calculations carried out in the lattice approximation, which will be discussed in more detail below, show that the attractive force between quarks is independent of their separation, so the potential energy is proportional to this distance. This results in the infrared confinement of quarks, which perhaps explains why free quarks and gluons are not observable.

A clear picture of confinement can be obtained by seeing what happens when a quark and antiquark forming a colorless bound state are pulled apart. When their separation is such that their potential energy exceeds the threshold for creating a new quark–antiquark pair, it becomes energetically more favorable to create such a pair and form two new colorless bound states, as shown in Fig. 31.

21. Nonperturbative methods in quantum chromodynamics

Perturbative calculations can be carried out when the coupling constant is small. As discussed above, asymptotic freedom ensures this only at short distances. Perturbation theory is not applicable in the region determining all of low energy hadron physics. In particular, perturbation theory cannot help elucidate the mechanism providing quark confinement, the hadron mass spectrum, and several other fundamental questions. It therefore becomes necessary to develop methods for carrying out QCD calculations without assuming that the interaction is small.

The $1/N_c$ expansion

The basic idea of the $1/N_c$ expansion is the following. In the new perturbation theory where the expansion parameter is $1/N_c$ (N_c is the number of colors, equal to three in the real world) one attempts to calculate low energy effects which are not amenable to the usual perturbative analysis.

The zeroth-order approximation corresponds to the formal limit $N_c \to \infty$ in QCD. In this approximation, at energies close to the meson production thresholds it is possible to isolate the dependence on the color-singlet degrees of freedom corresponding to pseudoscalar mesons. Then in the functional integral expressions for the fundamental quantities of the theory the integration over the remaining color variables can be done explicitly. The resulting theory can be described by a certain nonlinear effective Lagrangian which turns out to coincide with a phenomenological meson Lagrangian proposed earlier for describing hadronic phenomena at low energies.

This effective Lagrangian has nontrivial topological solutions of the solitary wave type called solitons. The properties of these solutions allow them to be identified as baryons. Approximate calculations of various characteristics of the soliton solutions, such as the charge distribution and magnetic moments, agree with the corresponding baryon quantities measured experimentally to within about 30%.

Vacuum condensates and sum rules

The hypothesis of quark–hadron duality permits certain hadron characteristics to be related to the results of perturbative QCD calculations by means of sum rules at finite energies.

An important step in the development of nonperturbative techniques occurred when the sum rule method was supplemented by the idea that the quark and gluon fields could have nonzero vacuum expectation values, referred to as condensates. This was first suggested by A.I. Vaĭnshteĭn, V.I. Zakharov, and M.I. Shifman.

A quark "condensate" is defined as the vacuum expectation value of a product of quark field operators $\langle\bar{q}(x)q(x)\rangle_0$. In the usual approach based on the interaction representation and perturbation theory this expectation value is equal to zero. A nonzero quark condensate, like the analogous gluon condensate $\langle F^2_{\mu\nu}\rangle_0$, implies that the QCD vacuum is not physical "empty space," but that it contains a nonzero density of quark–antiquark pairs and gluons.

In this new approach the observable characteristics of hadrons and resonances are expressed in terms of condensate expectation values. The numerical values found for the latter are quite large:

$$\langle\bar{q}q\rangle_0 \approx -(240\ \text{MeV})^3, \quad \frac{9\alpha_s}{32\pi}\langle F^2_{\mu\nu}\rangle_0 \approx (240\ \text{MeV})^4.$$

Therefore, from the physical point of view nonperturbative effects in QCD are significant.

QCD on the lattice

A very important direct, quantum field method of study which naturally supplements perturbation theory is the lattice approach. Here the Monte Carlo method is used for the approximate numerical calculation of fundamental quantities in QCD represented as path integrals approximated by multiple integrals of high order. The lattice variant of QCD is a model simplification of local QCD and is obtained by the following sequence of approximations.

Figure 32. Lattice link on which the gluon field is defined in the lattice approximation of quantum chromodynamics.

Figure 33. Four links, forming a lattice plaquette, on which the gluon action is defined.

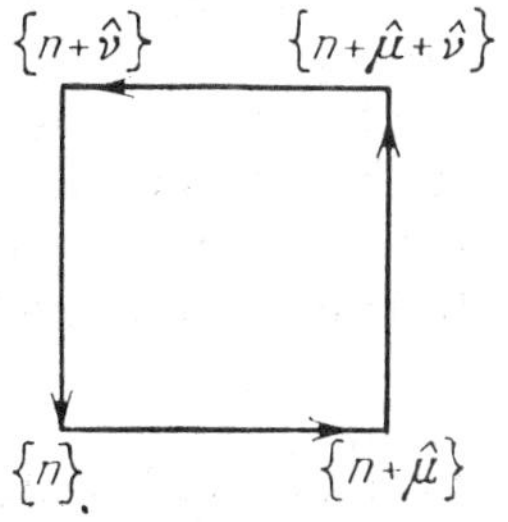

First, a rotation is made to Euclidean space, i.e., instead of the time t the imaginary coordinate ix_4 is used. Then continuous four-dimensional space is replaced by a discrete set of points, a lattice ($x \to \{n_1,n_2,n_3,n_4\} \equiv \{n\}$) with the same spacing a in all four directions.

Continuous fields therefore become functions of discrete variables. All derivatives are replaced by finite differences, and spatial integrations are replaced by sums over lattice sites. The quark fields are specified on the lattice sites, $\psi(x) \to \psi(\{n\})$, and the gluon fields are specified on the links: $B_\mu(x) \to u(\{n\},\hat{\mu})$, where $\hat{\mu}$ is a vector of length a giving the direction of the link (Fig. 32).

The action describing the quark–gluon interaction takes the form:

$$S(\psi,u) = -\frac{1}{2}\sum_{\{n\},\hat{\mu}} [\bar{\psi}(\{n\})\gamma_\mu u(\{n\},\hat{\mu})\psi(\{n+\hat{\mu}\}) \\ - \bar{\psi}(\{n+\hat{\mu}\})\gamma_\mu u(\{n+\hat{\mu}\},-\hat{\mu})\psi(\{n\})]. \qquad (21.1)$$

Therefore, quarks located at neighboring lattice sites interact via the gluon field defined on the link joining these sites.

The gluon action is somewhat more complicated. It contains all possible ordered products of gluon fields corresponding to the four links belonging to the same plaquette (Fig. 33):

$$S(u) \sim \sum_{\text{plaquettes}} \mathrm{Tr}[u(\{n\},\hat{\mu})u(\{n+\hat{\mu}\},\hat{\nu}) \\ \times u(\{n+\hat{\mu}+\hat{\nu}\},\hat{\mu})u(\{n+\hat{\nu}\},-\hat{\nu})]. \qquad (21.2)$$

It is easy to show that in the limit $a \to 0$ expressions (21.1) and (21.2) become the usual action of QCD.

The introduction of a lattice with finite spacing a also plays a role in the regularization of the ultraviolet divergences associated with the local description. In order to eliminate the divergences arising from the infinite extent of the lattice, a lattice of finite size ($0 \leqslant n_1, n_2, n_3, n_4 \leqslant L$) with opposite boundaries identified is usually used.

After the 4-dimensional spatial lattice is introduced, the functional integrals of QCD can be written as multiple integrals of high order. The most direct and universal method for computing them is the Monte Carlo method. The order of an integral over a gauge field of the group SU(N) is $4L^4(N^2-1)$. The maximum lattice size $L_{\max}$ for which such numerical integration can be performed is determined by the speed and memory capacity of the available computer and at present is a number of order 10.

Lattice calculations have shown that, for what would seem to be a small number of points ($L \sim 10$), the continuum limit is approached "sufficiently closely" for such small lattices to be used to obtain fairly reliable qualitative and partially quantitative physical results. The most important are the following.

It has been shown that the Coulomb potential between two static quarks is replaced at large distances by a linearly growing potential $U(r) \sim kr$. This dependence ensures quark confinement. It has thus been established that confinement in QCD is a consequence of the dynamics associated with the nonlinearity of the gluon field equations.

The first physical parameter calculated by the lattice Monte Carlo technique was the QCD scale parameter Λ introduced above. The value obtained was 200 MeV, in agreement with the estimate obtained by comparing perturbative QCD to experiment (see Sec. 20).

The masses of several hadrons (for example, the A_1 and ρ mesons, the nucleon, and the Δ_{33} resonance), decay constants, and the values of the quark and gluon condensates have been calculated. The agreement with experiment is of order 10%.

The masses of glueballs, colorless gluon structures which have not yet been seen experimentally, have been predicted.

The deconfinement temperature T_c at which a phase transition from the phase of hadronic matter with quark confinement to the phase of a quark–gluon plasma should occur has been calculated. The results indicate that in QCD $T_c \approx 200$ MeV, and for $T > 2T_c$ the temperature dependence of the energy density is given by the Stefan–Boltzmann law. This prediction is very important for cosmological models. It can be checked in experiments on collisions of relativistic nuclei, which are planned for the near future.

Although these quantitative calculations still have fairly large relative errors ($\sim 10\%$ and higher), the overall picture strongly indicates that both high energy strong interaction processes with large momentum transfer and the low energy, even static, properties of hadrons can be described on the basis of the QCD Lagrangian in terms of a small number of parameters characterizing the quarks and their interactions with gluons.

We should also note that recently there has been great interest in using the lattice approximation to analyze the interaction between gauge and scalar fields. This arose from the need to study the Higgs sector of the Glashow–Weinberg–Salam electroweak interaction model, grand unification models (see Sec. 23 below), and also supersymmetric theories (Sec. 24). Calculations using the lattice approximation of the Glashow–Weinberg–Salam model predict that the ratio of the Higgs boson mass to the *W*-boson mass is about 6 (with a 10% error). This result could be extremely important for experimental searches for Higgs bosons.

Chapter 7 Hypothetical models of interactions

22. The idea of the unification of interactions

The standard model

In the two preceding chapters we became acquainted with the present-day quantum field models of the electroweak and strong interactions. Both of these models are based on the principle of local gauge invariance, which allows the interaction carriers to be identified as the quanta of the corresponding vector fields. To study quantum chromodynamics and the Glashow–Weinberg–Salam model together, it is convenient to introduce the gauge group $SU_c(3) \times SU(2) \times U(1)$. This corresponds to the direct product of the gauge groups of these models and thereby formally unites the strong and electroweak interactions into a single model, the so-called standard model. Of course, here there is not any actual unification of the interactions, since they are associated with independent gauge transformations. This is expressed mathematically by introducing the concept of the direct product of groups. [We recall (see Sec. 3) that the transformations associated with each factor of a direct product of groups are independent and associated with different degrees of freedom.]

Interactions are characterized by three different coupling constants α_3, α_2, and α_1 corresponding to the groups involved in the direct product (instead of α_1 the usual quantum electrodynamical coupling constant $\alpha = \alpha_1 \cos^2\theta_W$ can be used).

As discussed above, the standard model contains three generations of fermions (quarks and leptons): (u,d,e^-,ν_e), (c, s,μ^-,ν_μ), (t,b,τ^-,ν_τ), and 12 vector particles (gauge field quanta): 8 gluons, the $W^\pm$ and Z bosons, and the photon. In addition, there are the as yet undiscovered massive scalar Higgs bosons associated with the SU(2) symmetry breaking.

When discussing the standard model, whose predictions agree with all the experimental data obtained so far, we should mention that it contains rather a large number of arbitrary parameters (the coupling constants, the masses of the quarks, leptons, and Higgs boson, and so on) which are not determined by the theory itself. The relation between the strong and electroweak interactions is also not specified by the model; it can be found only in a model which really unites these interactions.

The higher degree of symmetry at higher energies

Real progress in the construction of a unified theory of the strong and electroweak interactions came with the development of so-called grand unification models. These models are based on the hypothesis that at superhigh energies nature manifests a high degree of symmetry (so that the different types of particle nearly become identical). It is assumed that at these energies particles interact via a single interaction. The degree of symmetry in the organization of matter decreases at lower energies, and the single interaction "splits" into three "branches": the strong, weak, and electromagnetic, which manifest differing properties.

This hypothesis is actually quite natural, since in the known examples of broken symmetry the symmetry is generally restored at higher energies. One example is isotopic symmetry, which in pion–nucleon physics is a good approximation for kinetic energies much greater than the mass difference of the particles in question.

Effective coupling constants in the standard model

The starting point for grand unification models was the behavior of the effective coupling constants of the standard model. As explained in Sec. 3.4, when radiative corrections are included, the parameter determining the strength of the interaction in any process is not the coupling constant appearing in the Lagrangian, but some function of the momentum transfer k (or its conjugate length), which is called the effective coupling constant. Since in the units customarily used energy and momentum have the same dimension, in the standard jargon the variable k is referred to as an energy. It turns out that the effective coupling constants in non-Abelian gauge theories fall off with increasing k^2, while in other theories they increase as

$$\alpha(k^2) = \alpha(k_0^2\left(1 + \beta\alpha(k_0^2)\ln\frac{k^2}{k_0^2}\right)^{-1},$$

where $\alpha(k_0^2)$ is the value of $\alpha(k^2)$ determined experimentally at some energy k_0 [see Eqs. (13.6) and (13.7)].

The coefficient β determining the rate at which the effective constant changes depends on the number of fields participating in the interaction in question. When there are not too many matter fields in non-Abelian gauge theories it is positive. In particular, $\beta > 0$ for α_2 and α_3 in the standard model containing the particles known at present. If we assume that there are no other particles in nature except the known particles, whose masses are all less than 100 GeV, the energy (distance) dependence of the effective constants of the standard model has the form shown in Fig. 34.

The hypothetical unification of the effective coupling constants

The effective constants α_2 and α_3 decrease with increasing energy, while α_1 increases. It turns out that for energies near 10^{15} GeV (or distances of about 10^{-29} cm) all three effective constants become approximately equal.

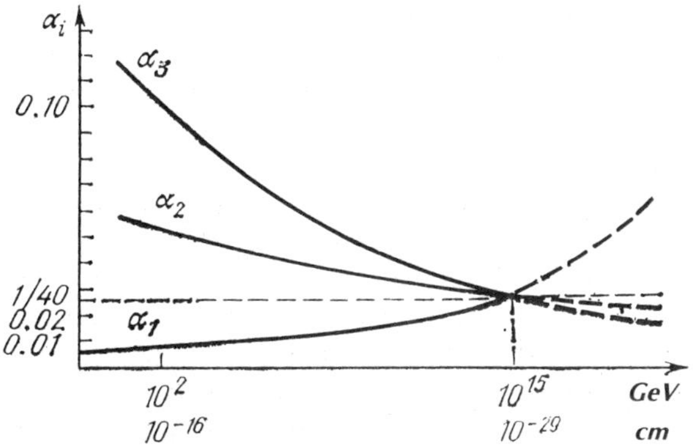

Figure 34. Behavior of the effective coupling constants in the standard model.

The fact that the interactions are comparable in strength suggests that the physical nature of the strong and electroweak interactions is the same, and that at distances less than 10^{-29} cm there exists a unified interaction with a single coupling constant. Graphically, this means that the curves do not diverge for $E > 10^{-15}$ GeV, as shown by the dashed lines in Fig. 34, but beginning at an energy $\sim 10^{15}$ GeV they merge into a single curve, as shown in Fig. 35.

A new symmetry and its breakdown

This unified interaction is based on a symmetry, of higher degree than the previous local symmetry, under transformations forming a group G, which possesses the group SU(3)×SU(2)×U(1) as a subgroup. The breakdown of this symmetry at distances exceeding 10^{-29} cm causes the single interaction constant to split into three: α_1, α_2, and α_3, each varying according to its own law.

The condition that the interactions be described by a single coupling constant amounts to the condition that the gauge group G be a so-called simple group. Roughly speaking, this means that G is not a direct product of other groups. In connection with this we note that the nature of the interaction unification we are discussing here is completely different from that in the Glashow–Weinberg–Salam model. In the latter the electromagnetic and weak interactions are not unified into a single one, but are simply mixed. The two coupling constants g_1 and g_2 remain distinct, which corresponds to the direct product of the groups SU(2)×U(1).

Therefore, the extrapolation of the behavior of the effective coupling constants of the standard model from energies $\lesssim 10^2$ GeV (or distances $\gtrsim 10^{-16}$ cm) accessible in current experiments, to energies of order 10^{15} GeV (distances $\sim 10^{-29}$ cm) and the assumption that a new symmetry exists lead to the following picture. At superhigh energies (very small distances), where

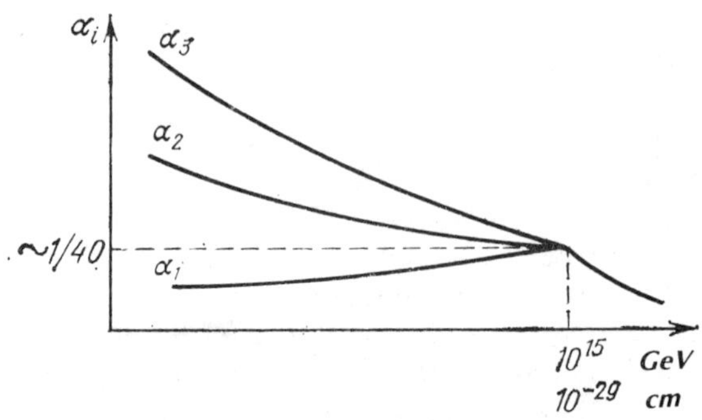

Figure 35. Illustration of the hypothetical unification of the interactions at energies above 10^{15} GeV.

the masses of all particles are insignificant, there exists a single interaction described by a gauge theory based on the group G. The first masses which become important as the distance is increased are of order 10^{15} GeV. They break the symmetry under the group G down to the symmetry of the standard model SU(3)×SU(2)×U(1), which in turn is broken down to SU(3)×U_{em}(1) at distances $\gtrsim 10^{-16}$ cm:

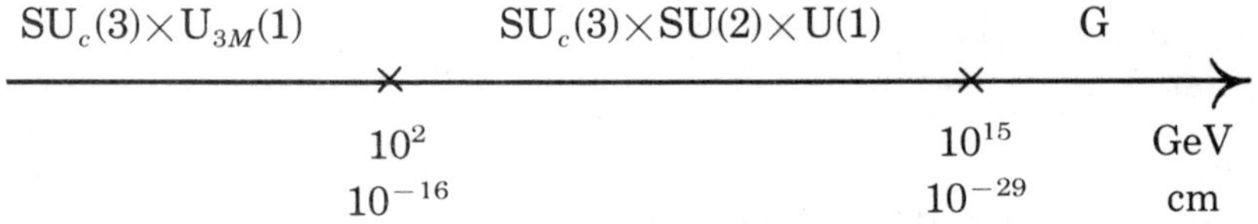

In the simplest variant of this model it is assumed that no new physical phenomena occur at distances from 10^{-16} to 10^{-29} cm, and, in particular, that there are no new particles with masses in the range from 10^2 to 10^{15} GeV. This range is therefore often called the "gauge desert."

In more complicated models, where it is assumed that the symmetry at superhigh energies is broken down to the symmetry of the standard model in several stages, additional particles with masses in this range can appear at each stage.

23. Grand unification models

In order to specify a particular grand unification model, we must fix the symmetry group G at small distances. The minimal group containing SU(3)×SU(2)×U(1) is SU(5). In addition to the SU(5) model, a number of other models based on different groups and having more complicated symmetry breaking schemes have been proposed.

Although the simplest variant of the SU(5) model is apparently ruled out by recent experimental data, as will be discussed below, we shall use this model as an illustration, because it is relatively simple and many of its features appear in some guise in other models.

The structure of the SU(5) model

In the SU(5) model each fermion generation is treated independently. The fermions are grouped into SU(5) multiplets such that leptons and quarks are contained in the same multiplet. This grouping of quarks and leptons into the same multiplet causes their electric charges to be multiples of each other: $Q_d = \frac{1}{3}Q_{e^-}$, $Q_u = -2Q_d$, and so on. This leads naturally to the observed equality of the absolute values of the proton and electron electric charges.

Let us make the following observation before turning to the description of the structure of the fermion sector. The field function of a fermion with nonzero mass can be split into parts with left- and right-handed polarization by using the projection operators $(1-\gamma_5)$ and $(1+\gamma_5)$, respectively. Under charge conjugation a left-handed fermion becomes a right-handed antifermion and vice versa. Therefore, instead of describing a fermion by a field function with two parts (left- and right-handed), we can use the equivalent description in terms of two field functions having a single polarization. A massless fermion has only one polarization direction. For example, the neutrino, which is assumed to be massless, has left-handed polarization. (We shall not discuss the possibility of the neutrino having a mass, since the experimental evidence is not yet convincing.)

The simplest fermion multiplet, a quintuplet (5-plet), contains the three left-handed antiquarks $(d_i^*)_L (i = 1,2,3)$, the left-handed electron $(e^-)_L$, and the left-handed electron neutrino ν_e (here the star will denote an antiquark):

$$\Psi^{(5)} = \begin{pmatrix} d_1^* \\ d_2^* \\ d_3^* \\ e^- \\ \nu_e \end{pmatrix}_L .$$

Another multiplet, a decuplet, contains ten left-handed fermions: the three quarks $(d_i)_L$, three quarks $(u_i)_L$, three antiquarks $(u_i^*)_L$, and also the positron $(e^+)_L$. The decuplet is conveniently written as a 5×5 antisymmetric matrix:

$$\Psi^{(10)} = \begin{pmatrix} 0 & u_3^* & -u_2^* & -u_1 & -d_1 \\ -u_3^* & 0 & u_1^* & -u_2 & -d_2 \\ u_2^* & -u_1^* & 0 & -u_3 & -d_3 \\ u_1 & u_2 & u_3 & 0 & -e^+ \\ d_1 & d_2 & d_3 & e^+ & 0 \end{pmatrix}_L .$$

The fermions we have listed together with their antiparticles make up the first generation.

Of the 24 ($= 5^2 - 1$) vector gauge fields, 12 are the familiar vector fields of the standard model: the gluons, the W and Z bosons, and the photon. The other 12, denoted by X and Y, are the carriers of the new interaction. Each of these has a color quantum number taking three values and fractional electric charge equal to 4/3 and 1/3, respectively. Omitting the numerical factors, the gauge field matrix can be represented schematically as

$$\hat{B} = \left(\begin{array}{ccc|cc} G;\gamma;Z & G & G & \bar{X}^1 & \bar{Y}^1 \\ G & G;\gamma;Z & G & \bar{X}^2 & \bar{Y}^2 \\ G & G & \gamma;Z & \bar{X}^3 & \bar{Y}^3 \\ \hline X_1 & X_2 & X_3 & \gamma;Z & W^+ \\ Y_1 & Y_2 & Y_3 & W^- & Z \end{array}\right).$$

If in the matrix $\hat{B}$ the elements corresponding to X and Y were zero, this would correspond to the standard model.

The interaction Lagrangian contains terms of the form $\overline{\Psi}^{(5)}\hat{B}\Psi^{(5)}$ and $\mathrm{Tr}(\overline{\Psi}^{(10)}\hat{B}\Psi^{(10)})$. It is easy to see that the first term leads to a direct interaction between quarks and leptons, described by vertices like those shown in Fig. 36. The second term contains, in addition to vertices for quark and lepton interactions with X and Y bosons, vertices in which only quarks interact with X and Y bosons (Fig. 37).

The vector bosons X and Y, which mediate transitions of quarks and leptons into each other, are called leptoquarks. Their masses are related to the energy scale at which the SU(5) symmetry is broken and are about 10^{15} GeV. This explains why the standard model is a good approximation at the energies presently attainable. The fact that these masses are so large makes the direct observation of leptoquarks impossible in the foreseeable future. However, effects due to the quark–lepton interaction can in principle be observed even now.

Proton decay

The most important consequence of this interaction is the existence of processes with baryon and lepton number nonconservation. The clearest example of such a process is proton decay, which can be represented as in Figs. 38 and 39. This decay can also occur via other subprocesses, including Y-boson exchange.

In order to see if this effect is consistent with our perception of the stability of matter, let us estimate the probability for this decay or, equivalently, the proton lifetime. The amplitude for this process is proportional to the square of the coupling constant and the propagator of the X (Y) boson:

$$\mathscr{A} \sim \frac{\alpha(M_X)}{M_X^2 - k^2} \approx \alpha(M_X)/M_X^2.$$

Figure 36. Vertex for a quark–lepton interaction with an *X* or *Y* boson.

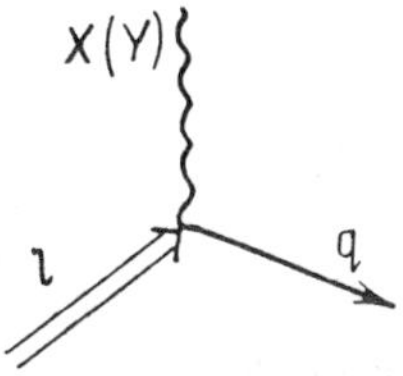

Figure 37. One of the vertices for the interaction of a quark with an *X* or *Y* boson.

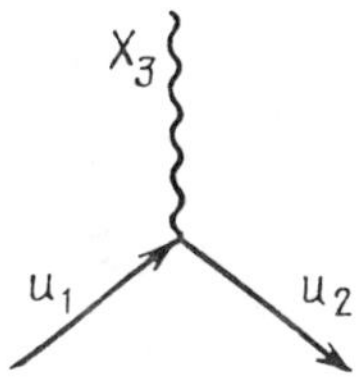

The proton lifetime is $\tau_p \sim 1/|\mathscr{A}|^2 \approx M_X^4/\alpha^2(M_X)$. The values $\alpha(M_X) \approx 1/40$ (see Fig. 34) and $M_X \sim 10^{15}$ GeV give $\tau_p \sim 10^{31 \pm 2}$ yr. (For comparison, we recall that the age of the Universe is about 10^{10} yr.)

Of course, this large lifetime makes it impossible to observe the decay of a particular proton. However, for a sufficiently large number of protons ($\sim 10^{32}$) the probability for at least one of them to decay can be significant. This is the principle upon which the current experiments to study proton stability are based. The results that have been obtained in such experiments so far rule out the simplest SU(5) model. In particular, the SU(5) model prediction $\tau(p \to e^+\pi^0) < 10^{31}$ yr is inconsistent with the experimental limit $\tau(p \to e^+\pi^0) > 10^{32}$ yr. It is not yet possible to experimentally check other grand unification schemes predicting a larger value of the proton lifetime.

Neutron–antineutron oscillations

New phenomena absent in the SU(5) model can arise for more complicated grand unification models. For example, in the SU(5) model processes with baryon number violation necessarily also involve lepton number violation. However, here the difference $B - L$ is conserved (for example, in proton decay $p \to e^+\pi^0, p \to \bar{\nu}_e \pi^+$). Processes in which $B - L$ is not conserved are allowed in models with more complicated groups containing a sufficient number of particles.

From the viewpoint of experiment, neutron–antineutron oscillations in the vacuum are the most interesting. In such transitions the baryon number changes by two units and the lepton number is not changed. Model-dependent estimates predict that the period of such oscillations is $\tau_{n \leftrightarrow \bar{n}} \sim 10^6$–$10^8$ s, and they could in principle be observed in sufficiently intense beams of neutrons

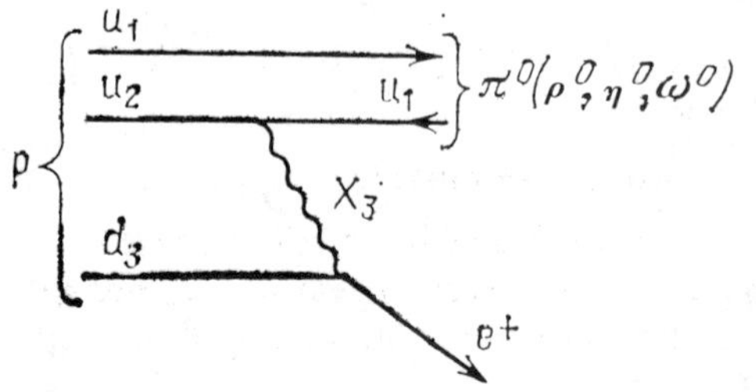

Figure 38. One of the diagrams describing proton decay.

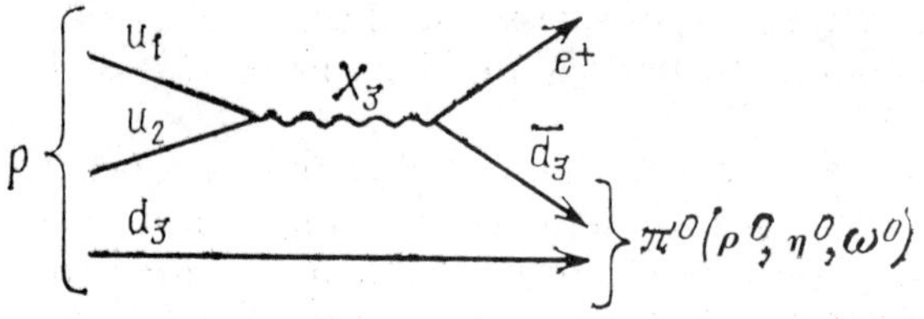

Figure 39. Another possible diagram for proton decay.

from reactors. However, so far there is no experimental indication that such processes exist.

Monopoles

In connection with the discussion of processes with baryon number violation predicted by grand unification models, we should mention another interesting phenomenon of these models. In the early 1930s Dirac noted that in quantum electrodynamics magnetic charges—monopoles—can exist without leading to any inconsistencies. In 1974 't Hooft and A.M. Polyakov showed that monopole solutions exist in gauge theories with semisimple symmetry groups which are spontaneously broken down to groups containing U(1). In particular, monopoles should be present in grand unification models, in which the symmetry group G is broken down to the standard model symmetry group $SU(3)\times SU(2)\times U(1)$.

The mass of such a monopole is related to the mass of the gauge boson X as $M_m \sim M_X/\alpha_G$, and for $M_X \sim 10^{15}$ GeV and $\alpha_G \sim 1/40$ it is a quantity of order 10^{16} GeV.

Without actually participating in an interaction itself, a monopole can induce processes with baryon number violation, in particular, proton decay, by acting as a sort of catalyst. V.A. Rubakov has shown that the electric and magnetic fields around a monopole create a fermion field condensate which violates baryon number with a probability proportional to the square of the fermion Compton wavelength $1/m_f^2$. Therefore, a typical cross section, for example, the cross section for the process:

$$p + \text{monopole} \to e^+ + \text{mesons} + \text{monopole},$$

must be comparable to the cross sections for ordinary hadronic processes.

It should be noted that up to now the various experimental searches for the monopoles themselves and proton decays induced by them have had negative results.

Limitations of grand unification models

Grand unification models and the wide variety of modifications of them are still speculative and in need of reliable experimental verification or refutation. There are also a number of purely theoretical problems associated with them.

The interaction unification we have discussed is not complete, since it does not include gravity. Attempts to include gravity in any unification scheme are not purely academic, since there always exists a region where gravitational effects become important. In fact, comparing the expressions for the potential energy of the gravitational attraction between two bodies of mass m, Gm^2/r, and the Coulomb interaction energy e^2/r, we easily find a dimensionless quantity analogous to the fine structure constant $\alpha = e^2/\hbar c \approx 1/137$. This quantity, $Gm^2/\hbar c$, characterizes the strength of the gravitational interaction. If a particle mass reaches the value $\sqrt{\hbar c/G} \approx 10^{19}$ GeV, the so-called Planck mass, this quantity is ~ 1. The distance corresponding to the Compton wavelength of such a particle, $r = \hbar/mc = \sqrt{G\hbar/c^3} \approx 10^{-33}$ cm, is called the Planck length.

Grand unification models are based on the rather artificial hypothesis that a "gauge desert" exists between the two characteristic energy scales differing by 13 orders of magnitude: the scale at which the symmetry group G is broken down to the group SU(3)×SU(2)×U(1) and the scale at which the group SU(2) in the electroweak interaction model is broken. It is not impossible that in this range, that is, at distances from 10^{-16} to 10^{-29} cm, particles manifest a deeper structure requiring new physical concepts.

The "population of the desert" by new particles occurs in many recent theoretical constructs. Of particular interest are models based on a new type of symmetry—the so-called supersymmetry, which will be discussed in the next section.

24. Supertheories

Supersymmetry

The progress in understanding the important role of internal symmetries in particle physics and the desire to discover their nature had led to numerous unsuccessful attempts to unify internal and spacetime symmetries in a nontrivial way. The failure of these attempts was made clear in 1976 when Coleman-Mandula theorem was proved. The theorem states that within the framework of relativistic field theory the internal symmetries generators B_i commute with the generator of spacetime translations, energy-momentum vector P_μ and the generators of Lorentz group $M_{\mu\nu}$:

$$[B_i, P_\mu] = [B_i, M_{\mu\nu}] = 0. \tag{24.1}$$

In order words, only trivial unification is possible when spacetime transformations have no effect on quantum numbers and internal transformations do not affect coordinates, that is, only the direct product of the Poincaré group with internal symmetry group.

As a consequence of this theorem, the internal symmetry generators commute with operators P^2 and W^2 ($W^\mu = -\frac{1}{2}\varepsilon^{\mu\nu\lambda\sigma}P_\nu M_{\lambda\sigma}$) whose eigenvalues determine mass and spin of the field, and therefore all members of a multiplet of the internal symmetry have the same mass and spin.

This theorem deals with only such internal symmetries that have generators forming Lie algebra. If one considers the so-called graded Lie algebras where instead of commutators of generators their anticommutators are defined, these symmetries are not forbidden by the theorem.

It was this symmetry, called later supersymmetry, that has been first proposed in 1971 by Yu.A. Gol'fand and E.P. Likhtman. Supersymmetry transformations make it possible to extend the Poincaré group in the nontrivial way by the inclusion of spinorial generators Q_i and $\overline{Q}_j$.

The commutators of these generators with P_μ and $M_{\mu\nu}$ are defined. And their anticommutator with each other is proportional to the energy-momentum operator:

$$\begin{aligned} &[Q_i, P_\mu] = [\overline{Q}_i, P_\mu] = 0, \\ &[Q_i, M_{\mu\nu}] = \tfrac{1}{2}\sigma_{\mu\nu}Q_i, \quad [\overline{Q}_i, M_{\mu\nu}] = -\tfrac{1}{2}\overline{Q}_i, \bar{\sigma}_{\mu\nu}, \\ &\{Q_i, \overline{Q}_j\} \equiv Q_i\overline{Q}_j + \overline{Q}_j Q_i = 2\delta_{ij}\sigma^\mu P_\mu. \end{aligned} \tag{24.2}$$

Here σ^μ are the set of Pauli matrices with $\sigma^0 = 1$. Antisymmetric tensors $\sigma_{\mu\nu}$ and $\bar{\sigma}_{\mu\nu}$ are also expressed in terms of Pauli matrices: $\sigma^{k0} = -\bar{\sigma}^{k0} = i\sigma^k$ ($k = 1,2,3$), $\sigma^{12} = \bar{\sigma}^{12} = \sigma^3$ and cyclic. Q_i and $\overline{Q}_j$ are two-component (Weyl) spinors. The index i ($i = 1,...,N$) indicates that more than one pair of such spinorial generators can exist. If $N > 1$, it is extended supersymmetry. If also $\{Q_i, Q_j\} = \{\overline{Q}_i, \overline{Q}_j\} = 0$, the supersymmetry algebra has no central charges.

As it follows from Eq. (24.2),

$$[Q_i, P^2] = 0, \quad [Q_i, W^2] \neq 0. \tag{24.3}$$

It means that supersymmetry transformations relate the states with the same mass but different spins (the action of each spinorial generator changes spin of a state by 1/2). Thus, a supersymmetry multiplet consists of fields with different spins but is degenerate in mass. Since these fields obey different statistics, supersymmetry is also reffered to as Fermi–Bose symmetry. The quanta of the fermionic and bosonic fields entering into the same supermultiplet are called "superpartners."

From the viewpoint of phenomenology, supersymmetry can exist only as a broken symmetry. If it were exact, the masses of bosons and fermions, the superpartners, would be the same and that certainly does not occur in nature.

None of the particles known so far is the superpartner of another. Therefore this scheme predicts that the total number of particles in nature is at least twice the number known at present. For example, in addition to the quarks and leptons, there exist so-called squarks and sleptons with spin 0, and in addition to the gluons and the photon there must exist the gluinos and the photino with spin 1/2, and so on.

Supersymmetric quantum field theory

The first supersymmetric quantum field model has been constructed by D.V. Volkov and V.P. Akulov in 1972. There was a nonlinear realization of supersymmetry, and the model was nonrenormalizable. A year later J. Wess and B. Zumino presented a renormalizable supersymmetric quantum field theoretical model of interaction of a spin 1/2 field with two spin 0 fields. And since then supersymmetric models became popular.

A compact and very useful technique was invented by A. Salam and J. Strathdee. They extended the usual spacetime by the inclusion of some anticommuting variables as additional coordinates.

The coordinates of this manifold, termed superspace, are four spacetime coordinates x_μ and an even number of two-component spinors θ^i, $\bar{\theta}^j$ (i, $j = 1,\ldots,N$) whose components are not ordinary numbers but form a Grassman algebra, i.e., they anticommutate with each other and are nilpotent (their squares are equal to zero). The simplest Grassman algebra (corresponding to the case $N = 1$) contains two two-component generators θ and $\bar{\theta}$ whose components satisfy the following relations:

$$\theta_1^2 = \theta_2^2 = \bar{\theta}_2^2 = \bar{\theta}_1^2 = 0, \quad \theta_1\theta_2 = -\theta_2\theta_1, \quad \theta_1\bar{\theta}_2 = -\bar{\theta}_2\theta_1, \ldots .$$

Supersymmetry transformations are written as transformations of the superspace coordinates:

$$x^\mu \to x^\mu + i\varepsilon\sigma^\mu\bar{\theta} - i\theta\sigma^\mu\bar{\varepsilon}, \quad \theta \to \theta + \varepsilon, \quad \bar{\theta} \to \bar{\theta} + \bar{\varepsilon}, \tag{24.4}$$

where ε, $\bar{\varepsilon}$ are two-component Grassman spinors.

Supersymmetry representations are realized by functions of the variables x, θ, and $\bar{\theta}$, called superfields. Superfields can be expressed linearly in terms of a finite number of ordinary quantum fields of differing tensor structure. In particular, one of the simplest, the so-called chiral superfield $\Phi(x,\theta)$ is written as

$$\Phi(x,\theta) = \varphi(x) + \bar{\psi}(x)\theta + \theta_1\theta_2 F(x), \tag{24.5}$$

where the component fields φ and F are scalar fields, and the component field ψ is a spinorial one. Higher powers of the Grassman coordinates do not enter owing to the nilpotency condition.

The formulation of quantum superfield theory on superspace is based on the rules of mathematical analysis on Grassman manifolds and especially the rules of integration over Grassman variables first defined by F.A. Beresin.

The action of superfield system is written in the form of the integral over the superspace, the supersymmetric Lagrangian being expressed in terms of superfields. For example, the kinetic action for chiral superfield is

$$S_0 = \int d^4x d^2\theta d^2\bar{\theta}\bar{\Phi}\Phi.$$

It contains the kinetic actions for component fields.

The supersymmetric Lagrangian of a system of interacting fields can be obtained by raising a superfield to some power, e.g., writing $g\,\Phi^3$, where Φ is the chiral superfield (24.5). We see that this gives rise to various interactions between the scalar and spinor component fields, with these interactions characterized by the same interaction constant.

Supersymmetry puts very rigid restrictions on the possible structure of the divergences in the theory. Various types of the interactions of component fields lead to correlated divergences so that the nonsupersymmetric contributions to counterterms cancel each other and renormalized theory is also written in terms of superfields.

The most complete cancellation of the divergences occurs in supersymmetric theories which also possess gauge invariance under some non-Abelian group. The physical basis for the divergences cancellation in these theories is that contributions of gauge fields and matter fields to divergent quantities differ by a sign. We have already encountered such a situation in analyzing the ultraviolet behavior of the effective coupling constant in quantum chromodynamics [see, for example, Eqs. (13.7), (20.1), and (20.2)]. Supersymmetry causes the bosonic and fermionic contributions to be correlated. Therefore in some supersymmetric models the divergences cancel exactly leading to the finite theories, not only in first order, but also in all other orders.

Supergravity

The supersymmetry transformations we have discussed so far are global. However, it is also possible to incorporate the local supersymmetry transformations with the parameters $\varepsilon(x)$, $\bar{\varepsilon}(x)$. As it follows from Eq. (24.4), in that case the group contains also the general coordinate transformations. So, a supersymmetric generalization of the general theory of relativity, called supergravity, can be obtained in this way. This theory was first proposed in 1976 by D. Freedman, P. van Nieuwenhuizen, S. Ferrara, and also by S. Deser and B. Zumino. The theory was further developed by many physicists. Besides those mentioned above, there are B. deWit, A. van Proeyen, V.I. Oghievetskii, E.S. Fradkin and others.

The transformation of the gauge field corresponding to local supersymmetry must contain the term $\partial_\mu \varepsilon$. Hence spin of this field is equal to 3/2. Its superpartner, the gravitational field with spin equal to 2, is the gauge field for local transformations of spacetime coordinates.

Thus, the simplest supergravity multiplet contains two particles: the graviton with spin 2 transmitting the gravitational interaction, and gravitino, a hypothetical neutral particle with spin 3/2. More complex supergravity multiplets contain also gauge vector fields corresponding to other interactions, and also matter fields with spins 1/2 and 0.

The application of supergravity models to the description of the world of real particles is based on the hypothesis that all four interactions are unified at energies above 10^{19} GeV, i.e., at distances smaller than 10^{-33} cm.

Owing to the tensor nature of the gravitational field and the nonlinearity of the interaction, the standard quantum theory of gravity is nonrenormalizable. Divergence cancellation in some theories possessing global supersymmetry gives rise to the hope that it might be possible to construct not only renormalizable but a finite supergravity model. Though finiteness has been proved for some supergravity models in the first two orders of perturbation theory, on the whole possible finiteness in all orders seems rather questionable.

The compactification of "extra" dimensions

Supergravity theories in four-dimensional spacetime, which contain a large set of fields, admit a natural formulation in spacetimes of higher dimension.

The idea of using spacetimes of dimension greater than four has quite a long history. In the 1920s T. Kaluza and Klein proposed a field theory in five-dimensional spacetime as a unified formalism for describing gravitation and electromagnetism. In their scheme the metric tensor of five-dimensional spacetime has a special form. It is expressed in terms of the components of the metric tensor $g_{\mu\nu}(x)$ of four-dimensional spacetime specifying the gravitational interaction and the vector potential of the electromagnetic field. In addition, they assumed that the extra fifth coordinate takes values on a circle S_1 of radius R ($0 \leqslant x_5 \leqslant 2\pi R$). The group of motion of this space is the group U(1) on which the electromagnetic interaction is based. In other.words, instead of ordinary 4-dimensional spacetime M_4, one has the five-dimensional spacetime corresponding to the direct product $M_4 \times S_1$. Therefore, here we encounter the simplest example of the compactification of an extra spatial dimension.

Kaluza-Klein theories represent one of the first attempts at the geometrical unification of spacetime and internal symmetries and the interactions associated with them. The contemporary theories are based on similar ideas about the introduction of extra spatial dimensions.

An essential feature of the extra coordinates is the fact that they must be compactified, so they can take values only in closed and bounded (compact) manifolds. The simplest examples of such manifolds are a circle and the surface of a sphere.

Although the introduction of extra dimensions had only a formal mathematical origin, it is not impossible that these dimensions have a deeper physical meaning. In order to understand how the presence of extra spatial coordinates can be related to the observed four-dimensional spacetime, let us consider the example in Fig. 40 of the compactification of an infinite plane in one of its two coordinates. Here the plane is transformed into the surface of a cylinder. The characteristic length is the cylinder radius R. At distances much smaller than R the curvature of the cylinder is practically unobservable and the surface appears flat. However, at distances much larger than R the compact range of variation of the coordinate x_2 appears to shrink to a point, and the surface seems to be a line, i.e., a one-dimensional manifold. In other words, at large distances, where the compactified coordinates are unobservable, the number of dimensions is effectively decreased.

In supergravity models formulated in spacetimes of higher dimension it is assumed that the extra spatial coordinates are compactified with radius equal to the Planck length $R_{\text{comp}} \sim 10^{-33}$ cm. At large distances ($R \gg R_{\text{comp}}$) the space is effectively 4-dimensional. Evidence that the "true" spacetime has more dimensions is that the 4-dimensional vectors and tensors have "extra" components, which are perceived as extra independent scalars, vectors, and so on. The extra coordinates begin to be noticeable at intermediate distances $R \sim R_{\text{comp}}$. Finally, for $R \ll R_{\text{comp}}$ all the coordinates play equal roles.

It has recently been suggested that at distances smaller than the Planck length not only is the number of spacetime dimensions increased, but also local quantum field theory is replaced by a supersymmetric theory of extended one-dimensional objects, the so-called superstrings.

Superstrings

The starting point of the theory is the model of the relativistic string, obtained by generalizing the relativistic dynamics of a point to extended one-dimensional objects. The relativistic dynamics of a point is determined by the action, which is proportional to the length of the world line described by the point in spacetime. A natural generalization is the assumption that the string action is proportional to the area of the world surface swept out by the string as it moves through spacetime. The principle of least action, which specifies the law of motion of the string in spacetime, corresponds from the geometrical point of view to finding the minimal surface in Minkowski space passing through the initial and final positions of the string.

The classical relativistic string possesses an infinite set of oscillatory modes. When quantized, these lead to an infinite series of states of the string with unboundedly increasing mass. The spacing of the masses in the spectrum is determined by the only dimensionful parameter—the string tension T, which has dimension $[m]^2$ (the string length $l \sim T^{-1/2}$).

This model was first suggested in 1970 by Y. Nambu in order to explain the Veneziano amplitude that expressed many features of hadron interactions. But some time later, J. Scherk and J. Schwarz remarked that it is more

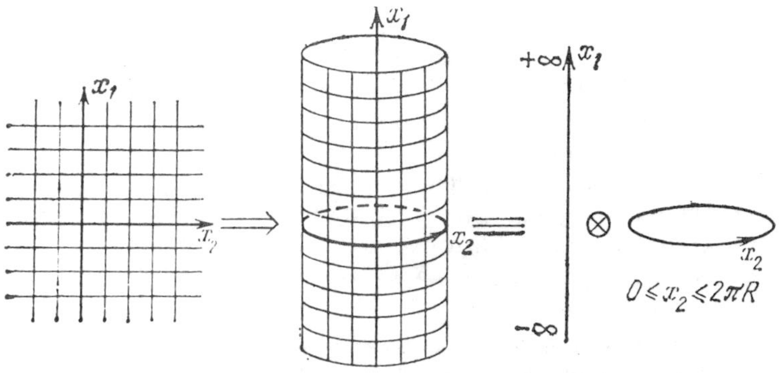

Figure 40. Example of the compactification of an infinite plane in one of its coordinates. The operation consists of transforming to a different range of variation of the coordinate x_2 which, instead of a infinite line, becomes a closed, bounded manifold—a circle. The plane is transformed into the surface of a cylinder, which can be represented as the direct product of a straight line and a circle.

natural to consider the string theory as a model of gravitational interaction. It means that the natural length scale for strings was the Planck length (10^{-33} cm) and $\sqrt{T} \sim 10^{19}$ GeV.

In 1971 P. Ramond, A. Neveu, and J. Schwarz succeeded in assigning a spin degree of freedom to every point of a string and thus obtained a new object, a spinning string.

The construction of the supersymmetric action for a spinning string leads to the so-called superstring. A consistent quantum theory of the superstring can be formulated only in ten-dimensional spacetime.

Although strings themselves are nonlocal objects, their interaction is local—it occurs at a single spacetime point. A single string can break into two, two strings can join at their ends to form a single string, and so on. An interaction between superstrings generates interactions between the states obtained when the strings are quantized. It turns out that the interaction between the ground (massless) states of superstrings in the local limit leads to supergravity and super-Yang–Mills theories in ten-dimensional space. In general, these limiting gauge theories contain both divergences and anomalies. In 1984 M. Green and J. Schwarz discovered that in some superstring theories with certain groups of gauge symmetry, the anomalies cancelled exactly. For these theories they also proved the cancellation of the divergences in the first order of string perturbation theory.

Although such superstring theories are formulated in ten spacetime dimensions with a huge symmetry group it is supposed that spontaneous compactification of the extra six dimensions and spontaneous symmetry breaking occur at the Planck energy scale.

The geometric and topological characteristics of the compact six-dimensional space, as it was made clear by E. Witten, determined, e.g., the group of symmetry remaining after spontaneous breaking, and the number of particle generations. E. Witten, with collaborators, proposed some types of manifolds as possible candidates for compact six-dimensional space and studied their phenomenological properties.

A new type of superstring, termed heterotic string, was suggested by D. Gross with co-workers. The main idea that forms the basis of the heterotic string construction is that the left-polarized and the right-polarized modes of the string are decoupled. Thus, there can be supersymmetry in the left sector only. Moreover, some new gauge invariance is introduced in the right sector in a certain nontrivial way.

Further development of this approach results in construction of a great variety of heterotic strings in four dimensions directly. Now more work to study the phenomenological predictions of these models is needed.

Thus, now it is hoped that in the framework of superstring theory it will be possible to formulate the true unified theory which at distances smaller than the Planck scale would be finite, and with increasing distance scale, would give a realistic description of all four fundamental interactions by a local quantum field theory. But the final answer of these hopes can be obtained only in the future.

25. An overview of quantum field theory

In the preceding pages we have attempted to give a reasonably coherent, though possibly sketchy, panorama of the more than half-century long history of the development of quantum field theory and its current state.

The nature of the evolution of quantum field theory

Several general conclusions can be made after analyzing the time evolution of this subject. As noted above in Chap. 2, the basic ideas and techniques of contemporary quantum field theory had for the most part been formulated by the mid-1950s. This was about the time when several factors, of both experimental and theoretical origin, led to doubts about the value of quantum field theory as the basis for particle theory and its popularity began to significantly decline among a large number of physicists. This process intensified during the following decade, which was dominated by the appearance of semiphenomenological constructs such as Regge pole theory, the analytic S-matrix, unitary symmetry, current algebra, quark bags, and other schemes of an even more pragmatic nature, with no deep relation to field theory.

However, during this "time of troubles" certain deep, barely observable developments were occurring in theoretical physics, which, owing to their lack of relevance to immediate physical needs, were officially classified as "theoretical theory." These developments led to fundamental advances: first, the construction of the quantum theory of gauge fields, and then by the late 1960s and early 70s they bore brilliant fruit in the form of the theory of the electroweak interaction and quantum chromodynamics.

From the conceptual point of view the important result was the discovery of a simple and universal method of constructing the interaction dynamics of relativistic quantum fields based on the principle of gauge symmetry. The gauge mechanism greatly simplified the logical foundation of the theory of interacting fields. However, quite a high price had to be paid for this simplicity.

The explicit formulation of the gauge dynamics, which is intrinsically related to the group nature of the symmetry transformations, makes extensive use of the language of the theory of continuous groups. In addition, the development of the formalism for quantizing gauge fields led to wider use of the functional integral representation. Because of this, the mathematical apparatus of the theory of Lie groups and also functional analysis techniques have by now become common tools in particle theory.

Meanwhile, the simplicity of the gauge mechanism is based on rather abstract physical concepts. Here there are two quite different components. One is the general motivation for local gauge transformations, which appeals to the unobservability of the phase of a complex quantum field. The second is the unobservability of the color degree of freedom forming the basis of the gauge symmetry leading to quantum chromodynamics, and also the unobservability of the quanta of its fundamental fields—quarks and gluons.

We can therefore point to two trends. One, which in the present context is fundamental, is the systematic simplification of the logical scheme of particle theory, the decrease of the number of independent fundamental physical assumptions and parameters, accompanied by the rapid growth of the number of observed facts and regularities (including the discovery of new particles, quantum numbers, and approximate symmetries). The second, associated trend (which is largely responsible for hindering the spread of new ideas and their comprehension by the wider physics community) is toward increasing complexity of the mathematical techniques which are used and a greater degree of physical abstraction. Let us consider these last two factors separately.

The increasing complexity of the mathematical techniques

The increasing level of mathematical complexity is a characteristic feature of theoretical physics throughout its entire history. The leading edge of physical theory has always made use of new mathematical methods, and in some cases has even stimulated their development in mathematics itself. Examples of physical theories for which this has happened in the recent past are the theory of electromagnetism, the kinetic theory of matter, the theory of relativity, and quantum mechanics. Quantum field theory has made extensive use of generalized functions, the theory of continuous groups, and the functional integral. As if in a relay race, quantum field theory rapidly "grabs" new branches of mathematics, and then immediately "passes" newly created theoretical techniques along to other areas of physical theory.

A familiar example of this is the Feynman diagram technique together with the associated functional integral representation. Already in the 1950s it was used in quantum statistics, and then later it found application in several other areas of theoretical physics for studying systems with a large number of degrees of freedom.

Another illustration is the renormalization group method. Discovered in the mid-1950s in quantum field theory, 30 years later it had become a standard technique in the theory of critical phenomena, the physics of polymers, theories of wetting and fracturing, the theory of turbulence, nonlinear transfer theory, and several other areas.

Yet another example is the practice of using the latest fast computers for carrying out algebraic transformations and analytic and symbolic calculations. This area of computer applications, which came into existence in about the middle of this century, was being used successfully by the 1970s to analytically calculate Feynman diagrams in higher orders of perturbation theory, for which special programs were written. Later, largely owing to the efforts of field theorists, systems of programs were written for analytic calculations of a more general nature. These universal systems found diverse applications in various areas of theoretical physics, mechanics, and mathematics.

It can therefore be stated that quantum field theory plays a pioneering role in the introduction and dissemination of the newest mathematical "technology" in the exact sciences and to a large degree sets the mathematical level in the present-day natural sciences.

Abstraction and visualizability

As noted above, in the last quarter century quantum field theory has come to use fundamental physical concepts and quantities of a new type which are not directly observable. Quantum theory is becoming more and more abstract. Not only is an electron trajectory unobservable, but so are the quantum number "color" on which the strong interaction dynamics is based and the quanta of the fundamental fields. Moreover, it is now usual to regard the vacuum as a state having little in common with the physical vacuum. It can be said that in the theory of the microworld the level of physical abstraction is now higher than that in quantum mechanics. As is apparent from the preceding section, this tendency to use concepts and images ever more remote from those in "everyday" physics is becoming stronger. Therefore, the theory of particles and their interactions, which lies at the leading edge of contemporary physics, is a source for ever more complex and abstract representations of the structure of the physical world.

This would seem to lead to the conclusion that the physics of the microworld is becoming less and less visualizable. In fact, it is becoming ever farther removed from the ideas and images of everyday human experience. It should, however, be realized that in our age of rapid scientific and technical progress the very concept of visualizability is not rigid. This is true both for mankind as a whole, and for any specialized groups of people. For example, because of the successes of space exploration and the advanced technology of information transmission, the idea of the Earth as a sphere is no longer an abstraction for an ordinary person in the street in the second half of our century. The fundamentals of quantum mechanics are taught in elementary physics courses and, in any case, are well known to a wide circle of scientific intellectuals, who make up a significant fraction of the population. The class of ordinary, so to speak, "semivisualizable" concepts is constantly being enlarged by newer and newer abstractions which first appeared in the natural sciences.

The quantum theory of subatomic particles and their interactions contributes to this progress by generating new and ever more complex ideas and images, which gradually enter the consciousness of physicists of other specializations and eventually reach a wider audience.

Recommended Literature

1. B. A. Arbuzov and A. A. Logunov, "Structure of elementary particles and relationships between the different forces of nature," Sov. Phys. Usp. **20**, 956 (1978).
2. B. A. Arbuzov and A. A. Logunov, "Particles and forces: Searches for unity," in *Science and Humanity* [in Russian] (Znanie, Moscow, 1979), p. 153.
3. V. S. Berezinskiĭ, "Unified gauge theories and the unstable proton," Priroda, No. 11, 24 (1983) [in Russian].
4. N. N. Bogolyubov and D. V. Shirkov, "The renormalization group? It's very simple," Priroda, No. 8, 3 (1984) [in Russian].
5. D. Z. Freedman and P. van Nieuwenhuizen, "The hidden dimensions of spacetime," Sci. Am. **252**(3), 74 (March 1985).
6. S. L. Glashow, "Quarks with color and flavor," Sci. Am. **233**(4), 38 (October 1975).
7. S. L. Glashow, "Toward a unified theory: Threads in a tapestry" (1979 Nobel Lecture in Physics), Rev. Mod. Phys. **52**, 539 (1980).
8. S. Hawking, "Is the end in sight for theoretical physics?," CERN Courier, No. 2, 73 (March 1981).
9. V. A. Matveev, R. M. Muradyan, and A. N. Tavkhelidze, "On the path toward uncovering the structure of elementary particles," in *Science and Humanity* [in Russian] (Znanie, Moscow, 1980), p. 1979.
10. Y. Nambu, *Quarks* (World Scientific, Singapore, 1985).
11. *The Nature of Matter*, edited by J. H. Mulvey (Oxford University, New York, 1981).
12. L. B. Okun', *Particle Physics* (Harwood, Chur, 1985) [Russian original, Nauka, Moscow, 1984], Vols. 1 and 2.
13. "The physics of the microworld," in *Small Encyclopedia* [in Russian], D. V. Shirkov, chief ed. (Soviet Encyclopedia, Moscow, 1980).
14. C. Quigg, "Elementary particles and forces," Sci. Am. **252**(4), 81 (April 1985).
15. L. H. Ryder, *Elementary Particles and Symmetries,* revised ed. (Gordon and Breach, New York, 1986).
16. A. Salam, "Gauge unification of the fundamental forces" (1979 Nobel Lecture in Physics), Rev. Mod. Phys. **52**, 525 (1980).
17. A. Salam, "Einstein's last idea: The unification of the fundamental interactions," Priroda, No. 1, 54 (1981) [in Russian].
18. D. V. Shirkov, "The development of the fundamental ideas of quantum physics," Priroda, No. 7, 11 (1979) [in Russian].
19. D. V. Shirkov, "Quantum symmetries in particle interactions," Sov. J. Part. Nucl. **14**, 445 (1983).
20. M. A. Smondyrev, "Intermediate vector bosons," Priroda, No. 12, 21 (1983) [in Russian].
21. M. A. Smondyrev, *Quantum Electrodynamics and Experiment* [in Russian] (Znanie, Moscow, 1984).
22. G. 't Hooft, "Gauge theories of the forces between elementary particles," Sci. Am. **242**(6), 104 (June 1980).

23. *Toward a Unified Field Theory (A collection of articles)* [in Russian] (Znanie, Moscow, 1980).

24. S. Weinberg, "Conceptual foundations of the unified theory of weak and electromagnetic interactions" (1979 Nobel Lecture in Physics), Rev. Mod. Phys. **52**, 515 (1980).

25. V. F. Weisskopf, "The development of field theory in the last 50 years," Phys. Today **34**(11), 69 (November 1981).

26. G. Wentzel, "Quantum field theory before 1947," in *Theoretical Physics of the Twentieth Century* [in Russian] (IL, Moscow, 1962).

The following books are recommended for more systematic study of quantum field theory and particle theory (the asterisk * indicates more difficult texts):

*A. I. Akhiezer and V. B. Berestetskiĭ, *Quantum Electrodynamics*, transl. of 2nd Russian ed. (Interscience, New York, 1965) [Russian original, 4th ed., Nauka, Moscow, 1981].

*V. B. Berestetskiĭ, E. M. Lifshits, and L. P. Pitaevskiĭ, *Quantum Electrodynamics*, 2nd ed. (Pergamon, Oxford, 1982) [Russian original, Nauka, Moscow, 1980].

J. D. Bjorken and S. D. Drell, *Relativistic Quantum Fields* (McGraw–Hill, New York, 1965).

N. N. Bogolyubov and D. V. Shirkov, *Quantum Fields* (Benjamin/Cummings, Reading, MA, 1982) [Russian original, Nauka, Moscow, 1980].

*N. N. Bogolyubov and D. V. Shirkov, *Introduction to the Theory of Quantized Fields*, 3rd ed. (Wiley, New York, 1980) [Russian original, 4th ed., Nauka, Moscow, 1984].

*L. D. Faddeev and A. A. Slavnov, *Gauge Fields: Introduction to Quantum Theory*, transl. of 1st Russian ed. (Benjamin/Cummings, Reading, MA, 1980) [Russian original, Nauka, Moscow, 1978].

R. P. Feynman, *The Theory of Fundamental Processes* (Benjamin, New York, 1961).

*C. Itzykson and J.-B. Zuber, *Quantum Field Theory* (McGraw–Hill, New York, 1980).

Nguyen Van Hieu, *Fundamentals of the Method of Second Quantization* [in Russian] (Energoatomizdat, Moscow, 1984).

*L. B. Okun', *Leptons and Quarks* (North–Holland, Amsterdam, 1982) [Russian original, Nauka, Moscow, 1981].

L. B. Okun', *Introduction to Gauge Theories: Texts of Lectures* [in Russian] (Moscow Engineering Physics Institute, Moscow, 1984).

J. P. Ramond, *Field Theory: A Modern Primer* (Benjamin/Cummings, Reading, MA, 1981).